Best Person Rural

BEST
PERSON
RURAL

Essays of a Sometime Farmer by

NOEL PERRIN

Selected, with a foreword, by

TERRY OSBORNE

DAVID R. GODINE · *Publisher*

BOSTON

For Ned

First published in 2006 by
David R. Godine, Publisher
Post Office Box 450
Jaffrey, New Hampshire 03452
www.godine.com

LIBRARY OF CONGRESS CATALOGING-IN-PUBLICATION DATA
Perrin, Noel.
Best person rural : essays of a sometime farmer / Noel Perrin ;
selected, with an introduction, by Terry S. Osborne. — 1st ed.
p.   cm.
ISBN 1-56792-307-0
1. Country life—Vermont—Anecdotes.
2. Vermont—Anecdotes. 3. Perrin, Noel. I. Title.
S521.5.V5P45 2006
974.3—dc22
2006002954

FIRST EDITION
Printed in Canada

Foreword

IN HIS FOREWORD to *Third Person Rural* Ned Perrin proclaimed that collection would be his last. Eight years later, in his Foreword to *Last Person Rural*, he admitted that his life trajectory had twisted and turned in ways he had not expected. With his marriage to Anne Lindbergh he was back to farming, and suddenly there was inspiration for another book.

He made no comparable proclamation in *Last Person Rural*. Maybe he thought the title of the book was obvious enough; maybe he thought there was the remote chance of another rural book. Whatever the reason, he thought better of predicting the future.

Good thing, too. Not because there were any surprises in his writing; he continued producing essays, articles and book reviews. But his life kept twisting: Anne died suddenly in 1993; his granddaughter Alexandra was born in 1996; he married Sara Coburn in 2001; and that same year he was diagnosed with Shy-Drager syndrome, a debilitating Parkinsonian-like condition that led to his death in 2004.

For those of us left in the wake of this final surprise, something seemed unfinished. A silence unfilled. There was a sense that Ned's life needed a closing, a final rural proclamation, a collection that would recognize it was indeed the last, and make its deliberate farewell.

But a "best of"? How would Ned have felt about that? Actually, that's easy. Ned loved "best of" lists. He barely made it through a day without rating something in terms of something else. Food was his favorite: the best tomatoes, potatoes, corn, red flannel hash, waffles, syrups, pies. I'm not sure he ever wrote about them, but he had mental lists of the best Vermont country diners and church suppers. That was only the beginning, though. He rated electric cars and pickup trucks and chainsaws and fencing and fenceposts. He had his favorite breeds of cows and sheep. He rated stone walls and wallbuilding techniques. And he wrote a widely read piece on the most environmentally committed colleges in America. So, though he might have been hesitant to have his own rural pieces collected into a "best of" list, he certainly would have understood the impulse.

How does one go about choosing the best of Noel Perrin's rural essays? I've been at it for months, and I'm still not sure. There are so many wonderful pieces. For a long time I kept changing my mind about which ones to include. In fact, if there weren't a deadline for this book, I'd probably still be changing my mind. To help me, I decided to use another of Ned's favorite tactics: I took an unscientific poll. Ned loved taking polls; for instance, in the *First Person Rural* essay "Market Research in the General Store," he and his daughter Amy set up a free syrup-tasting table in the Thetford Center Village Store to learn people's preferences on syrup grades and brands. He did this kind of thing all the time. For my poll I called and emailed people around Vermont and New Hampshire, asking them to name their favorite rural essays of Ned's. And here are the results: no one essay received a clear majority. In fact, no single essay received more than one vote. No kidding. There was no overlap in the responses. Each person had his or her own favorite.*

That's pretty amazing. But it corroborated what I felt in reading Ned's essays repeatedly and making a preliminary list of

---

* In the spirit of full disclosure, here's a confession: just before the writing of this Foreword, and long after I'd stopped collecting data and had already chosen the current collection, a person who had responded months earlier mentioned in passing that he also loved the piece "Selling Firewood in New York," from *First Person Rural*. That essay had

favorites: for every one I chose, there were five or six similar ones I easily could have picked. And so the poll results were helpful; they eased the pressure of having to make the one "right" list. In such an embarrassment of riches, I could never include everyone's favorite; at the same time, each one I chose would very likely be *someone's* favorite. In a way it would be hard to go wrong.

Wouldn't it?

The result is this collection, which combines essays from Ned's four *Person Rural* books and a sampling of rural essays written after *Last Person Rural*. In selecting and arranging the essays, I've been guided by two goals. First, I wanted to represent as many kinds of Ned's rural essays as possible. Ned wrote about rural life using a number of different formats, from practical how-to's to broad philosophical commentaries. The pleasure of reading his work comes in part from this variety. It seemed important to represent that.

But it wasn't easy to do. Even Ned had trouble. In the Forewords to three of the four *Person Rural* books he tried to categorize his own essays (more of his lists), but the categories changed from one collection to the next. Using as many of his ideas as possible, I tried to create categories that would fit all of the collections as well as the later, uncollected pieces.

Ultimately I settled on four large groups, each of which contained smaller subgroups: "Part-time Farming" essays (divided into the topics of food, tools and equipment, animals, and specific skills and activities); "Rural Living" essays (arranged geographically from Thetford, to Vermont as a whole, to New England, to rural life in general); "Political" essays; and "Book Reviews." I decided to omit book reviews from this collection because two excellent collections already exist: *A Reader's Delight* and *A Child's Delight*. So what you will find here is a mixture of "Part-time

---

been mentioned on another person's list too. So unofficially, I guess, "Selling Firewood" wins the poll, with two votes. But since the polls had been closed for quite some time, I can't in good conscience make the results official. Doing something like that would be like going back and making Al Gore President. Wait....

Foreword

Farming," "Rural Living," and "Political" essays, distributed as evenly as possible among the four *Person Rural* books and the later selections, and arranged so that they carry, as faithfully as possible, the trajectory of the thirty-year story they tell.

Which leads to my second goal for shaping the collection this way: to have the essays tell a story. The story of Ned. Not just of the forty years he spent at his Thetford Center farm, but of the man himself. That goal presents the biggest challenge of this project, because in his rural pieces Ned rarely discloses himself directly. In part that has to do with the purpose of the essays: though most of them contain him as their main character, few were written *about* him. And in part it has to do with his discomfort with what he called the "fishbowl" quality of country life, where everyone lives "perpetually in public." As a result he preferred to portray himself in the same way he described the New England landscape in his *Last Person Rural* essay, "The Soul of New England":

> The central truth about our landscape is that it's introverted. It's curled and coiled and full of turns and corners. Not open, not public; private and reserved. Most of the best views are little and hidden.

Later in the same essay Ned applied that idea to the landscape of his own farm, saying that what you could see of the farm from the road was just a fraction of all that was there. If you went behind the farmhouse, for instance, you'd find his best hayfield and his children's sledding hill.

These essays are my attempt to take you behind Ned's house. Once we get there, you can enjoy the view and read these essays with the pleasure of easy conversation they're meant to inspire. Or if you want, you can take your time, look more closely, and get to know Ned through what he says and does, marking the changes over time. This won't be as easy, but hopefully it will inspire a comparable pleasure.

If you read the book with this second perspective, be prepared:

what you notice may surprise you and may not make sense. From my own experience and from having spoken with other people, I know that there were a number of Neds, many of them wonderfully inconsistent. Sometimes his essays convey this. For instance, in "A Truck with Pull," he writes, "I'm the impatient type; I want that stuff out *now*." But in "My Farm is Safe Forever," also from *Last Person Rural*, he writes, "Mostly it's just the way I operate. Slowly. I'm the sort of person who can decide he needs a tractor and then spend several years thinking about it before actually going out and buying one." Both are true; both represent the Ned I knew. So, like his beloved New England landscape, the Ned you find here will be full of turns and corners. But all of it is him.

And all of it rings with passion, with a persistent love for the life he lived. That love animated every place in his life, from his farm, to the town of Thetford, to Vermont, to New England, to the rural world in general. It influenced his feeling for farming and farmers – from Floyd Dexter, his farmer-mentor-friend, to Ed and Ellis Paige and the other farmers with whom he interacted in the region, to the "bolt weevil" Minnesota farmers who in the 1970s resisted a cross-state power line with their own brand of ecosabotage. It shaped his relationships with his family and friends and students and readers. And it resolved a question asked in "Two Letters to Los Angeles," from *Third Person Rural*, a question so perplexing that Ned posed it periodically throughout all of his rural writing: "Why on earth does anyone choose to live here?"

"Love" was the answer he kept coming back to: a person lives here because he loves it here. But so fierce a love for a place, Ned well knew, came at a risk. He exposed that risk in "Vermont Silences" in *Second Person Rural*: "Never admit to caring too much about anything," he wrote, giving voice to the unsettling fear hidden behind rural stoicism, "because if you do, you'll probably lose it." Later in the book that same fear introduces "The Rural Immigration Law," which begins, "Each man kills the thing he loves...."

If there was a particular spot that taught Ned about these two impulses – an embracing love of place and the fear of losing it all

– it was his farm; and if there was an essay that best expressed what he had learned, it was "Farewell to a Farm." This last piece in the collection is the last he ever wrote, and is really more a love letter than an essay. ("I love every acre of it," he wrote of the farm.) The word "love" appears in the piece eight times and would have appeared many more, I suspect, if Ned had not chosen to replace it with related expressions such as "fond of," "dote on," "like," and "delight in," in order to avoid unwanted repetition. Despite such passion the essay seems to conclude that there is no way, finally, to avoid losing the things we love most.

From this perspective the story of Ned and the trajectory of his life on his farm seem to conclude with loss, and in some sense they do. But that would only be part of a more complicated ending. No good New England story, like the landscape in which it takes place, is simple; its ending always lies a little bit hidden, sometimes coiled inside itself, sometimes tucked in corners. To find it all, you might have to look in places you hadn't expected to. Here are two examples of those places:

🐌 His essay "My Farm is Safe Forever": in it he describes the way he donated to the town of Thetford the development rights to his farm, thereby ensuring that it "will stay a farm long after I have moved into the village cemetery." Against the nostalgic longing in "Farewell to a Thetford Farm," here is some unexpected hope – notably the wise recognition that one way to ensure you won't lose what you love is to give it away. Also implicit in the piece is the possibility – since "My Farm is Safe Forever" was written many years before "Farewell to a Thetford Farm" – that a person's final farewell may not necessarily be his last word.

🐌 The kitchen table in Ned's farmhouse: that's where I sit now, writing this Foreword, warmed by woodstoves, looking out of the bay window Ned called his "farmer's TV" at the seven-acre hayfield and sledding hill you can't see from the road. How I ended up here, owning the farm of a man I knew for almost twenty years – who taught me about sugaring and stone wall building and chainsawing, who influenced my classroom teaching and inspired my decision to write nonfiction, and whom I loved like a father – is one of those mysterious surprises that can bend anyone's life trajectory, in this case, my own. I can't possibly explain it, nor should I try here, since that's another story, and not what this book is about.

But my presence at the kitchen table is relevant to Ned's life in this way: it's a reminder that a person's story rarely ends with his death; more often it deposits into, mingles with, and is carried away by, the currents of other people's lives. In that sense we all inevitably become, in ways we can't predict, the legacies of those who have touched us.

One of the most quoted lines of Ned's essays is the concluding sentence of "Grooming Bill Hill" from *First Person Rural*. Having changed his fenceline to add eighteen acres of pasture on the top of Bill Hill (with the help of Floyd Dexter), Ned closes the essay with fatherly pride, noting, "It will be no bad legacy to leave." As I sit here looking out the bay window at the late-winter snow dusting the sledding hill to my right, then out the front window at Bill Hill to my left, and finally at his *Person Rural* books on the table before me, I see clearly now that eighteen acres was just the start of what he left for us.

No, Ned. No bad legacy at all.

Terry Osborne
Ned's Farm
February 2006

Contents

❦

LAST PERSON RURAL

UNCOLLECTED PIECES

First Person Rural

Jan Lincklaen's Vermont

IT IS SEPTEMBER 1791. A young Dutchman named Jan Lincklaen is riding horseback up the muddy road from Rutland to Burlington, Vermont. Once an officer in the Royal Dutch Navy, Lincklaen is now in the real estate business. He is the American scout for a giant land investment company back home in Amsterdam. The company already owns four million acres of land in New York State. Now it is thinking about buying 23,000 acres of maple sugar groves in Vermont. This is a pilot agricultural project. If all goes well (it's not going to), Dutch housewives will someday sweeten their coffee and frost their cakes with Vermont maple sugar, instead of cane sugar from the West Indies. Then the investors in Amsterdam can ease their consciences. They won't have to feel guilty about owning slaves to work the sugar plantations in Curaçao and Aruba.

The road Jan Lincklaen is riding along passes through frontier farm country. Most of it has been settled less than twenty years. Vermont is still so wild that in the very year he makes his visit, one upland farmer kills twenty-seven bears in a six-month period. It is so primitive that there is only one church bell in the entire state, way over at Newbury. There are no school bells at all, and not too many schools.

It is good farm country, though – boom country, like the Napa Valley a hundred years later. Jan Lincklaen likes what he sees. "The soil is very rich," he reports, and adds that it is particularly good for growing wheat and Indian corn. Up near Burlington, farmers are getting forty bushels of wheat to the acre, and up to seventy of corn. They are exporting beef to Canada.

There is wonderful hay land, too. Another visitor in the 1790s was dazzled to find farmers near Rutland who were making five hundred tons of hay a year – 20,000 bales, as we would say now. All this is from the virgin soil, which in another generation will be seriously depleted. Then farmers' sons from Vermont and New Hampshire will begin to stream out to the Middle West, to exploit a still richer soil.

But meanwhile northern New England is a place where a farmer can get rich. Nothing is commoner in those first thirty years of settlement, than to be able to buy a piece of forest for $1 an acre, clear it and get huge crops for a few years, and then sell out for ten or twenty times what you paid. Jan Lincklaen met a farmer in Dorset in 1791 who had just sold his sixty-acre farm for $19.25 an acre. In terms of present money, that would be something like $500 an acre – not too much less than land in Dorset brings right now.

The earliest farms would have looked very ugly to modern eyes. When a new farmer arrived, his first act was to chop down every tree on what was going to be his first field, cutting them about two feet from the ground. (This is a good chopping height.) He would then cut these giant oaks and hickories and maples into lengths, drag them into piles with oxen, and burn them. Then he would gather the ashes and boil them into pot ash, later called potash. Pot ash was the basic ingredient in eighteenth-century soap, and he could sell it for a high price. Or he could refine it still further into pearl ash, which people used then for baking powder. "These ashes amply pay them for the clearing of the land," a Vermont lawyer named John Graham wrote in 1795.

Well they might. Vermont pioneer farmers were producing about two million pounds of pot ash and pearl ash a year, and getting the then enormous price of 3¢ to 5¢ a pound. Shipped by water to New York or Philadelphia, barrels of ashes from Vermont and New Hampshire sold for a higher price per pound than tobacco, or flour, or even butter.

The pioneer farmer now had a stretch of rich virgin soil, dotted at frequent intervals with enormous stumps. His next step was to build a house. Here is a contemporary Vermont account of how he did it. (I have added a little punctuation.)

> When any person fixes upon a settlement in this part of the Country, with the assistance of one or two others he immediately sets about felling trees proper for the purpose. These are one to two feet in diameter and forty feet or upwards in length....
>
> The largest four are placed in a square form, upon a solid foundation of stone. This done, the logs are rolled upon blocks, one above another, until the square becomes about twenty or twenty-five feet high. The rafters are then made for the roof, which is covered with the bark taken off the trees.... The interstices in the body of the hut are filled up with mortar, made of the wild grass, chopped up and mixed with clay....
>
> In this manner is an abode finished, spacious enough to accommodate twelve or fifteen persons, and which often serves for as many years, till the lands are entirely cleared, and the settlers become sufficiently opulent to erect better houses. Three men will build one of these huts in six days.

Looking out from his windowless, chimneyless house onto a landscape of stumps, the pioneer farmer did not see desolation. Instead he saw visions of a glorious future. Here on that knoll would go the ten-room clapboard house which he would start to build as soon as the new sawmill in the village got going. There on the bottom land he would grow his hemp (for making rope, not drugs – his highs came from life), his flax, his wheat. As soon as the roots rotted, out would come all those stumps. With the oxen he would drag them into rows, and fence off some grazing land for the cattle. Meanwhile, it was time to plant an orchard, and to begin on some stone walls.

In most of rural Vermont and New Hampshire, these dreams came true in a hurry. It was only thirty-nine years from the settling of the first towns to the beginning of the nineteenth century. But men and women, three of whom can build a log house forty feet square and two stories high in a week can create a whole landscape in twenty years; and by the year 1800 the two states looked pretty much the way they do now, in their unspoiled sections, except that the soil and the people were both richer in 1800.

Not that life was easy. The famous rigor of our climate was the same then as now. A man who spent some time in Newfane, Vermont, in the 1790s complained, "This place is extremely cold and bleak in Winter, and not very hot in Summer." There were wolves in the mountains in such numbers as to make the keeping of sheep almost impossible.

But there was also abundance and prosperity – and an arcadian simplicity that in our own day seems almost incredible.

While he was inspecting Vermont farmland, Jan Lincklaen paid a call on the biggest farmer and second most famous man in the state. This was Thomas Chittenden, Captain General and Governor of Vermont. He was then rich in years and honors, not to mention land. He had become Vermont's first governor thirteen years earlier, in 1778, and he had governed uninterruptedly for eleven years. Then he stepped down for a year – and when the young Dutchman came to call in 1791 had just been triumphantly restored to office.

There was no Secret Service detail, or even a state trooper. There was just an old farmer. He showed the visitors into his house "without ceremony, in the country fashion." Lincklaen, who was used to admirals and twenty-one gun salutes, could hardly believe his eyes. "His house & way of living have nothing to distinguish them from those of any private individual, but he offers heartily a glass of Grog, potatoes, & bacon to anyone who wishes to come and see him."

Maybe rural Vermont is a little like that still.

[1978]

Grooming Bill Hill

QUESTION: Why is Vermont more beautiful than New Hampshire?
ANSWER: Because of Vermont farmers. Remove the farmers, and within ten years New Hampshire would surge ahead.

This is a serious argument. If you just consider natural endowment, the two states are both fortunate, but New Hampshire is more fortunate. It has taller mountains, it has a seacoast, it even owns the whole northern reach of the Connecticut River, except a little strip of mud on the Vermont side.

But New Hampshire's farmers mostly quit one to two generations ago and started running motels or selling real estate. The result is that most of New Hampshire is now scrub woods without views. Dotted, of course, with motels and real estate offices.

A lot of Vermont farmers, however, are holding on. Almost every farmer has cows, and almost every cow works night and day keeping the state beautiful. Valleys stay open and green, to contrast with the wooded hills behind them. Stone walls stay visible, because the cows eat right up to them. Hill pastures still have views, because the cows are up there meditatively chewing the brush, where no man with a tractor would dare to mow. (That's the other argument for butter besides its taste. I once figured that every pound of butter or gallon of milk someone buys means that another ten square yards of pasture is safe for another year.)

Until lately, my own contribution to the beauty of Vermont was modest. I did fence two little hayfields a few years ago, so that my neighbor Floyd Dexter could run beef cattle there after the hay was cut. Sometimes I run a couple myself. But both of these were good fields when I bought the place. My contribution was merely turning them from straight hayfield to hayfield-that-gets-grazed, so they would stop shrinking a little every year, and so that the cows would eat right up to the stone walls.

This year, however, I think I have seriously joined the ranks of those who maintain Vermont. Or maybe not so much joined as been quietly drafted by Floyd.

It all began because of Bill Hill. Bill Hill is a large lump of glacial debris behind the pasture across the road. I own it. Insofar as a thing as small as a human being can claim to own a thing as big as a hill.

Sixty years ago, it was all pasture. No trees except for one white birch on top, and a row of immense old maples on the slopes behind it. But just before World War II a New York lawyer bought this farm. He naturally kept no cows on Bill Hill. When I got it, one end was completely grown up to woods, and the rest was in every possible stage from briar-choked pasture to almost-woods. The top remained open, and because I like to picnic in a place with a 360-degree view, I have painfully kept it open by dragging a little sicklebar mowing machine up there every couple of years.

Last summer, though, I was watching Floyd's cattle uncover yet another stone wall in the field behind the house and trim the apple trees up perfectly to a height of five feet, and it struck me that there was a better way to maintain Bill Hill than dragging little machines up it. At that time my idea was just to fence four or five acres: the face of the hill we see best from the house, and the top.

The next day Floyd was over looking at a newborn calf, and I told him my idea. He liked it. Together we climbed Bill Hill and tentatively set the bounds. It turned out to be more like seven acres than five, because he pointed out that by using just a little more wire, I could include quite a lot more of the hill.

All winter when I had a spare afternoon I would go over and prune up bull pines and cut out poplars in the pasture-to-be, so as to encourage the grass. I got quite skillful at skiing out with a chain saw in one hand. Floyd got us a couple of hundred cedar posts at East Thetford Auction to supplement my remaining hemlock, and we bought wire at a remarkable store in Topsham called Freddy Miller's.

This spring, as soon as the frost was out of the ground, we began to drive posts. Also to enlarge the boundaries. The very first day we were out, Floyd led us as if by accident through a beautiful level patch of grass just beyond Bill Hill – and before I knew what happened I had agreed to fence nine acres instead of seven.

The boundaries stayed set for about a month. (We were fencing only on weekends, and not all of them.) Then one warm May afternoon, just as we were coming over the hill with the wire, almost ready to turn and close the pasture, Floyd remarked that it was thirsty weather. "I don't suppose there's any water back here," he said as we wiped our sweaty faces. I said no, not a drop.

We drove a few more staples in silence, and then Floyd remarked almost dreamily that he had gotten his feet wet deer-hunting behind the hill last fall. Probably dry there now, he added.

"I don't see that," I said. "If there was water there in November, there's certainly going to be in May. Let's go look."

Floyd was skeptical, but just to please me, he came. Sure enough, about two hundred yards beyond where I had meant to turn the fence, there was a good-sized wet place right near my boundary wall with Ed Paige, and even a tiny stream running. In thirteen years, I had never noticed it. Too grown up with briars and brush.

"Awful good to have water where you want the cattle to graze," Floyd said. "It'll keep them out on the hill. Course, this probably dries up along about June." As he spoke, he was walking steadily uphill from the wet spot to a place where someone had rocked in a spring, probably 150 years ago. People don't do that for places that dry up in June. We dug it out a little with our hands, let the water clear, and had a drink. I had never seen the spring, either. Floyd knows my land better than I know it myself.

Since the whole idea is to keep the cattle on the hill, I didn't even much resist taking the pasture on back, even though I had now committed myself to fifteen acres. And it was my own idea – Floyd wasn't even present – when I decided the next day to go back still further, to the stone wall by the maples, and turn the wire down that.

That's how I come to be adding eighteen acres of pasture this year. That's how come for the next half-century, at least, there will be one green grassy hill in Thetford Center, Vermont, to contrast with the dozen or so wooded ones, and a new green meadow behind it. There will be cows against the skyline, and there will be four new stone walls visible. It will be no bad legacy to leave.

[1977]

Sugaring on $15 a Year

MOST COUNTRY DWELLERS in New England sooner or later think about doing a little maple sugaring. About nine-tenths of them never actually get around to it. They don't have enough trees, or they don't have enough time, or they don't have the $700 that even a small evaporator costs. Retired people with time and maples and $700 generally don't have the stamina you need to keep slogging through the snow with full sap buckets.

If you are such a frustrated maple sugarer, I have a solution to offer whereby you can sugar this spring with no physical effort and for a total investment of roughly $15. In your spare time. Without setting one foot in the snow.

The trick, of course, is that I am using "sugaring" in its old and true sense – not to mean the production of maple syrup in an evaporator, but the production of maple sugar in a pot. This can be done starting with sap, of course, but it can also be done starting with existing syrup, any old syrup, which is the method I am proposing. Assuming you already have a kitchen with a stove in it, you need only three things to start sugaring: half a gallon of low-grade maple syrup ($7 or less); a rubber mold (about $8); and a touch of skill (free). With these simple ingredients you can turn out several pounds of really stunning maple candy.

Sometimes when people make the kind of claim I have just made – that, with no training and with practically no expenditure of time or money, you can do some wonderful thing – they are secretly expecting you to provide the wonderfulness yourself. I once bought a book on building stone walls, lured by a dust jacket which promised that with just the rocks lying around my fields, I could build handsome retaining walls, set stone steps in them, design beautiful stone culverts, and so on. All this was true. I could have – if I had a natural genius for setting stones. I don't. I can lay up serviceable stone walls, and after ten years of practice, that remains the limit of what I can do.

But making maple sugar is different. It really takes the merest touch of skill. You can be the sort who bends nails, even the sort who breaks the yolk without meaning to when you fry an egg, and still make maple candy that elicits actual moans of pleasure from those who eat it. Maybe not with your first batch, but certainly with your third or fourth.

Let me assume you are now convinced and dying to start. First, obviously, you've got to get some syrup and a mold. The mold you can pick up wherever maple sugaring supplies are sold, which means about one hardware store in three in northern New England. Or you can order one by mail from the Leader Evaporator Company in St. Albans, Vermont, or from the G. H. Grimm Company in Rutland. The most fashionable mold at the moment seems to be one that makes large and realistic maple leaves – I have one myself – but what I recommend for starting is a less complicated mold. Leader has one that makes fifteen hearts that I like pretty well, although I wish the hearts were smaller. (No child I know agrees.) It sells for $8.50.

As to the syrup, get the cheapest you can find. There is no *harm* in its being this year's Fancy Grade, but there's no need, either. Let me tell two stories to show just how low you can stoop.

The first involves my own introduction to making maple sugar. I got into it because of a small disaster. In the spring of 1975 I was making maple syrup as usual and managed to burn a batch of nearly two gallons just as I was about to draw it off. Since I only make about twenty-five gallons a year (I get tired, slogging through the snow with sap buckets), this meant goodbye to a sizable chunk of the year's production. Burnt maple syrup is not usable; you can filter out the thousands of flakes of black carbon, but you can't filter out the taste.

Fortunately, a friend named Tom Pinder came into the sugarhouse before I got around to throwing the stuff out – in fact, while I was still jumping up and down and swearing.

Tom can find a solution for almost anything. If he had been present at the burning of Gomorrah, he would either have figured out a way to put the fire out, or at least had the city rebuilt in no time. He took a look at my scorched pan, and then tasted the contents. "I've heard," he said thoughtfully, "that you can get the burned flavor out of syrup by taking it down to sugar." And he offered to help me try. I had never before even considered making sugar, because I thought it was too hard for someone of my limited skill.

As I've already suggested, it was easy. To my amazement the burnt flavor vanished almost completely, as did the dark color. We made pie tins full of sugar and cookie pans full of sugar, and finally we bought a mold and made maple-sugar leaves; our friends and families gobbled up what we didn't eat ourselves.

By early summer I had used up the whole two gallons, and I began looking around for a fresh supply. I recalled that back in 1974 another friend, Alice Lacey, had made six or seven gallons of dark Grade B late in the season and had never gotten around to canning it. It was still sitting in her mud room in a ten-gallon milk can. It sported a thick layer of green mold on top. It was now fifteen months old.

Alice was happy to trade me a gallon for a few fence posts, and when *that* syrup produced pale, delicious maple candy, I knew I was onto something. I promptly traded for three more gallons, and then Alice acquired a mold and turned the rest into sugar herself. We have both saved end-of-the-season syrup for sugar ever since.

The process we both use works like this. We pour about a pint (no need to measure) of syrup into a good-sized pot so that it's no more than an inch deep. Half an inch is better. By instinct rather than for any scientific reason I know of, I use a stainless steel pot rather than aluminum. Tom, who hears a lot, has heard that you should always use a wooden spoon rather than a metal one; since I have a wooden spoon, I do that, too. But here I'm not even prompted by instinct. I just like tradition.

You bring the syrup to a boil, and you then turn the heat well down, since even shallow syrup will boil over with extraordinary rapidity. In ten minutes or less you should have the right consistency for sugar. You can test with a candy thermometer or with a special maple-sugaring thermometer, if you like. Leader and Grimm both sell them. I find it simpler to use the hard-ball test: lift your wooden spoon and when the drops are coming off individually let one fall into a glass of cold water. If the sugar is ready, the drop will instantly form into a compact flattened ball in the bottom of the glass. If it isn't ready, the drop will spread out in a sort of little nebula.

You now take the pot off the stove and stir while it cools. To hasten the process, set it in cold water for thirty seconds or so, still stirring. With this shortcut, the sugar usually begins to crystallize within two or three minutes. But it is by no means cold; it is still too hot to touch.

The crystallizing is rather dramatic. What you have been stirring is a thick, opalescent brown taffy, incredibly sticky and tooth-pulling. It now begins to lighten in color and to get rapidly thicker still. Don't worry, just keep stirring. It will then abruptly set.

At this point, dash back to the stove (you have been stirring while comfortably seated at the kitchen table), and put the pot over medium heat. The still-hot sugar will re-liquefy within a minute or so, still keeping its crystalline structure. In fact, at this point the crystals usually get finer and the color a still paler and more elegant tan.

Keep the pot on the stove for another half minute after the sugar re-liquefies, stirring steadily. The sugar is now hot enough to stay pourable for several minutes, and you stroll back to the kitchen table and fill your mold. Since a pint of syrup makes almost a pound of sugar, you will have a little left in the pot as a starter for the next batch. So don't scrape or wash the pot; put it and the spoon away as is until the next time you sugar. Then add another pint of syrup and start boiling.

Meanwhile, you have fifteen maple-sugar hearts or twelve maple-sugar leaves, or whatever, sitting in your mold. You could even have letters spelling out the names of your children, since any supplier can get you an alphabet mold. Leave them there for ten or fifteen minutes so you won't burn your hands when you press them out. Then put one out to eat, still warm, and the others on a plate to cool. Be sure they get consumed within roughly the next two weeks, because after that the little cakes gradually dry out and get hard. I don't think you'll find this a problem. My difficulty has always been to keep everything from getting eaten the first day.

There are a good many variations I haven't gone into. Koreans are said to have twenty-some different names for cooked rice, depending on how much water they cook it with and hence how hard it is. Vermonters have only four names for different degrees of hardness in maple candy: maple cream (which is too soft for a mold), soft tub sugar, hard tub sugar, and cakes. But in actual fact, there are infinite gradations from soft tub to cake, and each tastes just a little bit different. All taste wonderful. This is why I predict that when you've run through your original half gallon of syrup, you'll be out scouring the countryside, looking for a gallon of Grade B to buy cheap.

[1976]

In Search of the Perfect Fence Post

❧

MIDWEST FARMERS – most of them, anyway – have a boring time with fence posts. When they need some more, they just open the Sears catalogue and order another five hundred metal ones. (Current price: $2.49 each.) Then they lay out one hundred or so in a straight line across the prairie, and start stringing wire.

New England is different. Within a mile of my house – well, two miles – I can look at fences strung on eleven different kinds of posts. Bud Palmer, who runs the garage in the village, has a horse pasture fenced with pine. Ellis Paige uses mostly split oak to restrain his Angus cattle. Barbara Duncan keeps her goats behind a mixture of oak and maple saplings. George deNagy uses hemlock for Push and Pull, his team of work ponies. Warren DeMont has metal posts. Not from Sears, but salvaged from a floodplain the Government took over a few years ago. Floyd Dexter, the best fencer of us all, uses entirely sharpened cedar posts. Except when he runs out in the middle of a job, that is, and then he's been known to use cherry, tamarack, lever wood ... almost anything except popple or elm. The town uses cut granite posts, six inches square and six feet long, as a good many farmers used to. My neighbor Dr. Lucius Nye has – but let me get on to my own story.

I've been fencing for sixteen years now. My serious fences surround three cow pastures totaling thirty-two acres. The posts are mostly cedar and butternut, decently soaked in oil.

But I also have a *jeu d'esprit*: a little half-acre sheep pasture (for two sheep) done entirely in green hemlock. Five young apple

16

trees fenced against deer with a variety of posts that could be called New England Miscellaneous. Plus more. Over the years I've probably put up two dozen fences of one sort or another. And since I started from a state of ignorance which farmers' sons usually pass beyond between the ages of five and six, I have made every mistake but one that it's possible to make. I've put up a fence without bracing the corners. Strung barbwire from the bottom strand up instead of the top strand down. Put the small end of a post in the ground instead of the large. The one thing I *haven't* done is to use white or gray birch for posts. And there it was poetry that saved me, not common sense. Long before I thought of being a farmer, I had read most of Robert Frost, and could quote from "Home Burial":

> *Three foggy mornings and one rainy day*
> *Will rot the best birch fence a man can build.*

It's not even much of an exaggeration.

Sixteen years ago I blundered into fence building when I acquired a wife and an old farm the same year. She was determined to have a garden, and the deer were determined she wasn't. I volunteered to build a fence.

Having all that newly acquired land, most of it covered with trees, I wasn't about to buy posts. I took my pulp saw and wandered around my new woods until I found a place where there were a lot of gray-barked young trees coming up three and four together. I know now they were sprout-growth elms and red maples. Even then I knew it was good forestry practice to take a tree coming up in three or four stems like that and cut it back to one main stem. So I sawed busily until I had twenty nice gray-barked posts, enough to fence the garden, six posts to a side. (If you think that would take twenty-four posts, you clearly haven't done much fencing. And if you retort that it would be twenty-four after all, because a big garden needs two gates and two posts for each, you'd be right. But I hadn't thought of that yet.)

The next step was to get them into the ground. Someone had told me that fence posts should be set two feet deep, so I went to one corner of the garden and dug as narrow a two-foot hole as my shovel would dig. Then I put one of the four largest posts in, shoveled the dirt back, and tamped it. Then I tested my new corner post, which easily moved through a twenty-degree arc. I spent the next half hour packing around it with rocks, and then retamping with another fence post rather than with my foot. The post now wiggled only slightly.

It was absurd to think of spending forty minutes per post, so I did what I usually do when I'm in a dilemma. I went to Dan and Whit's (in the opinion of many, one of the world's great general stores) to see what exotic tool they carried that would solve my problem.

Almost immediately I found a thing called a post-hole digger, and brought it home in triumph. It worked splendidly. The holes it made were a little too big for my posts, but only a little, and I could get fairly tight posts planted in about ten minutes each. No deer got into the garden that year.

Three years later, however, they could have shouldered their way in almost anywhere. Elm posts rot fast. Green elm posts rot faster. Green elm posts with the bark on rot fastest of all. If I were ever mad enough to use elm posts again, I would cut them in the spring when the bark is loosest, peel them, dry them for a year, and then soak them in used motor oil.

But I never have used them again. Not counting a small grazing area for pigs (ineptly electrified, and a total failure), my next fencing project was to make a paddock for a horse we had bought. This was the year that the first eleven garden posts rotted out; and overreacting as usual, I had decided to use posts that would *never* rot out. In fact, I got metal posts from Sears. The date was 1965, and back then the posts were a mere $1.29 each.

I am the sort of person who has three little helpings at dinner instead of one giant one, and this temperamental quirk carries over to buying farm equipment. Thus, though I figured the paddock would take two rolls of barbwire and seventy or eighty posts, what I actually bought was one roll and thirty posts. This was good luck, because I quickly discovered that however handy metal posts are

on the Great Plains, they aren't much use on the rollercoaster ter-
rain of a New England farm. You ask what about Warren DeMont
and *his* salvaged metal posts? He uses them on stock fence for *his*
two sheep a year. Stock fence goes on relatively level ground, and
they work fine. I'm talking about barbwire up and down hill. You
can't drive staples into a metal post, and it's nearly impossible to
get the wire tight.

So I switched in mid-fence to cedar. By this time, in my fourth
year with land of my own, I knew several of the farmers in town
and had walked land with them. I knew about cedar posts, and
about driving mauls. I had also learned to recognize most of the
common trees that grow in New England, and I knew that I personally had no cedars at all. No matter. They sell them at Agway.
In fact, I might have gone to Agway even if the farm had been one
big cedar swamp. The garden fence and the pig fence had been
such failures that I'm not sure I would have trusted a cedar I had
cut myself. I wanted something professional.

There were two gigantic piles of posts out behind the main
Agway building in White River Junction. All cedar, all sharpened
at the big end to a beautiful tapering point. Posts from one pile
cost $55 a hundred (Agway is getting $95 for the same posts now),
and posts from the other pile cost $45 a hundred. I am sorry to
report I did it again. I got posts from the $45 pile. It was my last
really *major* mistake in fencing.

Though any normal person enjoys saving ten dollars, it wasn't
for that that I got the cheap posts. I got them because they were
smaller. They ran 3″ to 4″ at the butt end, while the $55 posts were
4″ and up. I committed this idiocy out of fear. While I was vainly
trying to string wire on the metal posts, a farmer friend had given
me a couple of sharpened cedar posts so I could see what I was
missing. One was nearly 6″ in diameter (he said use it for a corner
post), and one was quite small. Even with my new twelve-pound
driving maul from Dan and Whit's, and my wife to hold, it had
taken me about forty full-strength blows to drive the big one. I
had to stop in the middle and rest. The last six or eight blows

pretty well splintered the top, and pretty well ruined me. By contrast, the smaller one had gone into the ground 2″ a whack, and left me feeling like the strong man at a county fair. So I concluded that big driving posts were for real farmers, and little ones were for transplanted urbanites like me.

This was, of course, a completely false conclusion. For which I paid the price six years later, when I had to redo the whole paddock because all my tiny posts were giving out. The true conclusion is this. You have to learn *how* to drive sharpened posts, and when you do, it's easy. You don't even need someone to hold the posts up for you (though it's more companionable that way).

Here is the true way to put up posts for a New England fence, learned from sixteen years of hard experience. Or, rather, here are two ways: the fast way and the best way.

The fast way is to go buy however many posts you need. Make sure you buy large ones, 4″ to 5″ in diameter at the butt, and sharpened so that the taper extends at least a foot. Throughout New England, such posts are normally cedar – though if you find locust, grab them. Meanwhile buy or borrow a soft iron driving maul – not to be confused with a sledge hammer, which has a much smaller head – and a good four-foot iron bar. Lay out your first strand of wire to make a line, or else do it with string. Then you can start driving posts. Take your bar, and drive it into the ground where you want the first post to go. If you hit a small stone, you can drive it on down with the bar. If you hit a big one, or ledge, move.

When you are down about twenty inches or so (usually about four easy whacks with the maul), stop. Then wiggle the bar around with a circular motion until you have a hole sort of like the inside of an ice cream cone, with the top about 4″ in diameter. Then pull the bar out, pace off the distance to the next post, and stick the bar into the ground where the post is to go. (That way you won't lose it.) Now go back and shove a post firmly in the first hole. It will easily stand, and you should be able to drive it in anywhere from six to a dozen blows, depending on what the soil is like. And it will be completely tight – no play at all.

In Search of the Perfect Fence Post

When you get to the big corner posts, you can do them the same way – but it won't be any six to twelve blows. If you have a post hole digger, now is when it's useful. It will make you a straight-sided hole about 5″ in diameter. Dig one a foot deep. Then work your bar for another foot in the bottom. Plump the 6″ corner post in, and drive it with ease.

That's all there is to the fast way. Provided, of course, you remember to brace all corner posts with braces that go right to the base of the nearest line post on either side. Little bitty braces do nothing. You can buy cedar poles for braces, or you can cut your own. Either way, they should be about a foot longer than the distance between your fence posts, and at least 3″ in diameter at the small end. Here's how to install them. With a shovel you make a small hole right up against a line post, and shove the butt end of the brace in. Then you lay the smaller end carefully on top of the corner post, and trim it to length with a neat diagonal cut. Then you slide it down the corner post for about a foot, which gets it good and tight, and nail it in place with a tenpenny (3″) nail. Rich people sometimes use sixteen or even twentypenny nails.

The best way is considerably more complicated, and also takes more equipment. But it's cheaper, and Robinson Crusoe would like it better.

First learn to recognize all the trees you have. If you don't already know how to chainsaw, learn to. Then start looking for stands of young trees that need thinning. In the absence of cedar, wild cherry or tamarack is best, though both hemlock and white pine will do. Don't bother with trees growing in the open; they taper too fast, and you'll get only one or two posts from a tree. Three to five is what you should be getting. Since you cut the posts six feet long, that means the tree should have eighteen to thirty feet of straight trunk before it gets too small. Cut a bunch.

Another option is to split out posts. Butternut probably works best. If you had some butternut trees nine to fifteen inches in diameter, growing straight and not too many lower limbs, you can make a lot of posts from one tree. You fell it, buck it into six-foot logs, and split each log into fourths or sixths. All it takes is two wedges, a sledge hammer, and a modicum of skill. Butternut is a notoriously weak wood, so you always make large posts. I myself wouldn't dream of sacrificing a good butternut tree just for posts – but where I wanted to thin a fenceline anyway, or where a butternut was growing too close to sugar maples I wanted to favor, I occasionally take one. I've probably made a hundred posts that way. Six trees' worth.

If you're smart, you will have cut all these posts where you can get pretty close to them with a pickup truck, which you now drive out there. Bring your wife (or husband, or unsuspecting houseguest) and an extra pair of ear protectors. Open the tailgate and load the first three or four posts in the back of the truck, with the butt end sticking out. The spouse or guest puts on the ear protectors and climbs in the back of the truck. While he or she holds the first post steady, you sharpen it with your chain saw. This amounts to cutting a slice off each side the full length of the chainsaw blade, getting the victim in the back of the truck to turn the post ninety degrees, and then cutting off two more slices. The whole procedure takes less than a minute. It leaves, incidentally, a pile of fluffy shavings like giant excelsior, which children find irresistible. Now do the other two or three, have the victim pile the sharpened posts on one side, and load the next batch.

If you're *really* smart, you will have done all this in the spring; and as each post is sharpened, you can also peel it. The four bark points where the sharpening cuts end will pull like Band-aids. I myself am rarely that smart and, to be honest, I think it matters only moderately.

By lunchtime you will have posts enough for a great deal of fence; and if you want to you can start pounding that afternoon. I have. If you're the deferred-pleasure type, though, instead bring them home and pile them under cover for a year. And the next spring, if you have the time, treat them. With a few posts, you just paint the bottom two and a half feet with creosote or whatever.

In Search of the Perfect Fence Post

But if you've cut a lot, that gets exceedingly boring – and besides, painting doesn't give deep penetration. The better way is to give them a twenty-four-hour soak. I have usually done this in an old fifty-five-gallon drum, in which I put a mixture of one-third creosote and two-thirds old motor oil. Say, five gallons of creosote and ten of motor oil. Three-thirds creosote would doubtless be better, but creosote is expensive. You can soak about fifteen posts at a time. I strongly advise tying the barrel to a tree, because when it's full of six-foot posts, it is top-heavy; and few things are more annoying than having a barrel filled with used motor oil and creosote tip over on top of you, or even not on top of you. The smell, my wife informs me, does not come out of work pants until the third washing.

Having done all this, you are ready to take your iron bar and your driving maul, and set posts. You will have a double satisfaction when you're done. You can look at your new fence and reflect that thanks to your skill it's going to last two and perhaps three times as long as ordinary fences. And (provided you take care not to count the purchase price of any mauls or chainsaws – which, after all, you still have, and will keep using), you can compute that your posts cost you about 10¢ each. That's chiefly for the creosote.

On the other hand, part of me hopes that you don't do all this – that you go and cut some basswood in the morning, and have it in the ground that afternoon. True, it will rot out as fast as my elm fence around the garden. But after three years it may be time to move a garden, anyway. And what else are young basswoods good for? Besides, you'll be contributing to that sense of variety which I hope New England never loses.

There are two Yale dropouts who are caretaking/renting a house about three miles from our village. They've made a big vegetable garden, and they have fenced it with chickenwire mounted entirely on alders. Skinny alders driven small end into the ground. What's more, it works. A deer could probably push that fence over with one hoof, but they don't. Nor do they jump over. I think they're as pleased and touched by that fence as I am, and wouldn't hurt it for the world.

[1977]

23

The Two Faces of Vermont

᛭

WHEN YOU CROSS the bridge from Lebanon, New Hampshire, to Hartford, Vermont, practically the first thing you see on the Vermont side is a large green and white sign. This bears two messages of almost equal prominence. The top one says, "Welcome to Vermont, Last Stand of the Yankees." The bottom one says, "Hartford Chamber of Commerce."

Only Vermont could have a sign like that, I think. Vermont makes a business of last stands. Consider just a few. It is the last stand of teams of horses that drag tanks of maple sap through the frosty snow. It is the last stand of farmers who plow with oxen and do the chores by lantern light. Together with New Hampshire and maybe a few places in Ohio, it is the last stand of dirt roads that people really live on, and of covered bridges that really bear traffic. It is the last stand of old-timers who lay up stone walls by hand, of weathered red barns with shingle roofs, of axmen who can cut a cord of stovewood in a morning – of, in short, a whole ancient and very appealing kind of rural life. This life is so appealing, in fact, that people will pay good money to see it being lived, which is where the trouble begins. There's a conflict of interest here.

On the one hand, it's to the interest of everyone in the tourist trade to keep Vermont (their motels, ski resorts, chambers of commerce, etc., excepted) as old-fashioned as possible. After all, it's weathered red barns with shingle roofs the tourists want to photograph, not concrete-block barns with sheet aluminum on top.

Ideally, from the tourist point of view, there should be a man and two boys inside, milking by hand, not a lot of milking machinery pumping directly into a bulk tank. Out back, someone should be turning a grindstone to sharpen an ax-making a last stand, so to speak, against the chainsaw.

On the other hand, the average farmer can hardly wait to modernize. He wants a bulk tank, a couple of arc lights, an automated silo, and a new aluminum roof. Or in a sense he wants these things. Actually, he may like last-stand farming as well as any tourist does, but he can't make a living at it. In my town it's often said that a generation ago a man could raise and educate three children on fifteen cows and still put a little money in the bank. Now his son can just barely keep going with forty cows. With fifteen cows, hand-milking was possible, and conceivably even economic; with forty you need all the machinery you can get. But the tourists don't want to hear it clank.

The result of this dilemma is that the public image of Vermont and its private reality seem to be rapidly diverging. My favorite example comes, of course, from the maple-sugar business. Suppose you buy a quart of syrup in the village store in South Strafford. It comes in a can with brightly colored pictures on it. These pictures show men carrying sap pails on yokes, sugarhouses with great stacks of logs outside, teams of horses, and all the rest. They are distinctly last-stand pictures.

But suppose you decide to go into the sugaring business for yourself. When you write away for advice, you get a go-modern or private-reality answer. You are told not to hang pails at all, much less carry them to the sugarhouse on a yoke. Instead, install pipes. Don't bother to cut any four-foot logs, you're told, even though your hills are covered with trees. Texas oil gives a better-controlled heat. And finally, your instructions say, the right way to market the stuff is to put it in cans that show men carrying sap pails, sugarhouses with great stacks of logs....

The state is full of this kind of thing. I've seen a storekeeper spend half an hour taking crackers out of plastic-sealed boxes and putting them in the barrel he thinks summer visitors expect him to have. I've driven over a fine old covered bridge, intact and complete from floor to roof, and just as busy with modern traffic as it ever was with wagons. Tourists stop constantly to get pictures. But should one of them go poking around underneath (it doesn't happen often), he would see that it secretly rests on new steel I-beams, set in concrete. The great wooden trusses up above are just decoration now.

Or take fairs. I've been at a fair where the oxen for the ox-pull were trucked in from as far as fifty miles away. The town was full of oxen. If you didn't happen to notice them arriving in the trucks, you'd have concluded that here was a real last-stand neighborhood. Or you would have until the pulling began. Then you might have gotten suspicious. What the teams were pulling was more concrete, big slabs of it. Furthermore, when each pair of oxen had made its lunge, a distinctly modern element appeared. This was a large yellow backhoe which would rumble up, belching diesel fumes, and give the slab a quick push back to the starting point. The net effect was rather like watching the Dartmouth crew at practice, which I've also done. The college boys are like the imported oxen. They use muscle-power. The crew surges up the river between Vermont and New Hampshire, every man pulling his oar for dear life. The coach is like the backhoe. He skims alongside in a fast motorboat, steering casually with one hand, and shouting orders through a megaphone he holds in the other.

Most of all, though, I see the difference between Vermont in photographs, in sentimental essays, in advertisements, and the state as it is actually getting to be. I'm thinking, for example, of roads. Even in California they know what a Vermont road is like. It's a last-stand road. It may be dirt or it may be blacktop, but what matters is that it's narrow and it follows the lay of the land. In most of Vermont, obviously, that means going in curves. The road will curve in so as not to spoil a field, curve out again afterwards, meander up a hill. It has, of course, a stone wall running along each side. Generally a row of big old trees marches beside each wall.

Often these are maples, and when they are, the farmer who owns them taps every spring, using buckets.

But what if some Californian gets sick of twelve-lane express-ways and moves to Vermont? What if he buys a house on such a road? He hardly gets the place remodeled (exterior unchanged, interior restored to authentic 1820, cellar packed with shiny new machinery) before the town road commissioner comes to see him.

The town's going to resurface the road next summer, the commissioner says. While the crew are at it, they plan to make a few other changes. They're going to take out all the sharp curves, reduce all the steep gradients, and widen the whole road by six feet. Twenty feet, if you count shoulders.

To the Californian's horror, it turns out that this will mean taking all the stone walls on one side, and most of the trees on both sides. It also turns out that the road will no longer follow the lay of the land. In particular, it's going to be raised four feet where it passes his house, and the road commissioner is hoping to use his stone wall for part of the fill. Next year the photographers will have to find some other road to put on their "Unspoiled Vermont" calendar. But two cars will now be able to pass in mid-winter without one stopping and the other slowing down to ten miles an hour. And reality and image will be a little further apart.

If the ex-Californian puts up a fight for his stones and his trees, he soon finds that the selectmen and the road commissioner are not wholly against him. They may think he shows his Los Angeles background in wanting to save a stone wall when it's barbed wire you need for keeping cows, but they don't really disapprove. In fact, the road commissioner freely admits to liking last-stand roads himself. He was raised on one. What's on everybody's mind, it turns out, is that the town is not going to get any State Aid unless it widens and straightens the road to state specifications. And, of course, a lot of people in town are tired of having to stop every winter when they see another car coming. But the money is the main thing. The commissioner rather thinks the state itself gets Federal road money on similar conditions. In other words, town and state are under the same pressure all the dairy farmers are: go modern or go broke. That's a strong pressure.

And yet it's not the only one. Opposed to it is the natural cussedness of Vermonters, lots of whom don't want to go modern. And some would say it's not just cussedness, either. There are deep satisfactions to last-stand life. And, finally, there is all that good money the tourists pay.

All this has amounted to almost equal pressures in the two directions, at least until very recently. Almost everyone in Vermont is at least partly on both sides. But most are more on one side than the other. By oversimplifying a little, one can draw up a sort of chart of the battle lines. In fact, I have.

Let me start inside the fort. Manning the loopholes, and actually making the last stand of the Yankees, are a hard core of hill farmers, country storekeepers, ox breeders, and so forth. Economically their pressure is small. Most of them earn less money every year. But they aren't about to quit. In my part of the state, a fair number have taken full-time jobs so that they can keep farming nights and weekends. These are the kind referred to on the sign.

Allied with them are about half the summer people. (The other half aren't opposed; they're neutral. In fact, they're mostly too busy water-skiing and playing golf even to have noticed that there *are* farmers in Vermont.) But the first half like coming to a region of old-fashioned farms, and having farmers for neighbors. They may not want to look after cows or lay up stone walls themselves, but they like to watch other people do it. Meanwhile, the money they pay out for care-taking, barn-painting, and meadow-mowing helps to keep a good many last-stand families going.

Also allied are nearly all the middle-class immigrants or so-called year-round summer people. Most of them were originally drawn here by last-stand life, and a certain number actually lead it. I know one couple, both with college degrees, whose first action on getting their Vermont farm was to disconnect the electricity. They do the chores by lantern light. I know another man, born and bred in Maryland, who has become as good a country plumber and as authentic a rural character as lives in New England.

Finally, there is a scattering of people outside the state who provide economic support in one way or another. Here are the covered-bridge lovers who send money to help a Vermont town keep one. The bridge I mentioned a while back drew contributions from no less than four covered-bridge clubs last year when the town it's in had to decide whether to repair it or to replace it with the latest thing in concrete bridges. Here also are the city people who will spend extra time and money to get locally made cheese, or barnyard eggs from hens raised organically, or hand-made wooden toys – and in doing so have put a good many country stores in the mail-order business. If you could only get that by mail, too, some of them would buy hill cider that's capable of turning hard, rather than the tame stuff (filtered, pasteurized, and practically castrated) that's available in supermarkets. If they only had trucks, some of the suburban ones would come up and buy half a ton of real manure for their gardens. The number of such people is small but growing.

Turning to the other side, an equally mixed group is pushing toward modernization. In the center are what I guess to be a majority of all native Vermonters under fifty, starting with the valley farmers who already have big herds and bulk tanks. They don't want to be the last stand of the Yankees. (After all, look where Custer was after *his* last stand.) They want their sons to be able to go on farming after them – even if the "farm" turns out to be a lot of hydroponic tanks inside a two-acre concrete shed, fronting on a twelve-lane expressway.

Nearly everyone concerned with either education or state government is also on this side, at least officially. So are all of us who drive to shopping centers instead of walking to the village store, who buy lumber at Grossman's instead of at the town sawmill. And so, with a superb irony, are many Vermonters in the tourist trade, plus the tourists themselves.

The irony is that the tourists don't know they are. They come here to look at last-stand life. They wouldn't cross the road to look at a supermarket or a two-acre concrete shed. Most of them firmly believe they're helping to support old-fashioned Vermont by coming here at all. But though they flock to see the last-stand country, and, if they're here in the spring, to take a free taste of hot maple syrup, or in the fall to do a little free hunting – free as far as the owner of the land is concerned: the state charges a stiff fee – inevitably where they spend most of their money is in the motels, filling stations, and restaurants. Last-standers get only a little directly. They don't get much indirectly, either. Even though the restaurant owner knows that his tourist customers have come to look at last-stand life, and even though he personally hopes it will survive, he's still in business. He mostly buys his eggs at the battery farm, his milk at the big automated dairy, his beef from Kansas City, and so on. His chief gesture toward last-standism is to make sure the syrup cans in his gift shop have pictures of sap buckets on them.

In the last five years the balance has perceptibly tipped in favor of modernization. Most people agree that the last stand is likely to end in about one more generation. What will happen then? Let me present an admittedly partisan view.

Most of Vermont will look like – well, it will look like central New Jersey with hills. Where there are now fields and meadows, there'll be scrub woods mixed with frequent tree plantations. Every now and then there'll be an automated concrete "farm." Around each lake will be a ranch-style summer resort. The entire state will be linked by superb highways. In the more rugged sections, these highways will take most of the valley land there is. (Right now a four-lane highway built to Federal interstate specifications consumes forty acres out of every square mile it goes through, or one-sixteenth of the whole square mile.)

There will, to be sure, be three or four villages left in which last-stand life goes on. Two of these, I guess, will be commercial ventures, and two will be owned by the state. All four will be pure fake. If you drove into one – I'm going to call it Old Newfane Village – first you'd see a wooden barn with four live cows in it, and a man specially trained to milk them. Then you'd notice a grove of maples next to an old-fashioned sugarhouse. Probably the maples will have to be made of plastic, with electric pumps inside, since the main tourist season begins in June rather than March, and since there's no way to keep a real maple from budding until June. But it will be genuine maple sap that the electric pumps draw up from a refrigerated tank under the sugarhouse.

Beyond the plastic maple grove will be a large woodshed. There, for 50¢ you'll be able to watch a man first sharpen his ax on a hand-turned grindstone and then chop up a couple of logs. Every twenty minutes he'll reblunt his ax by smacking it into a block of granite. An expert from Colonial Williamsburg will check his technique twice a day. And public image and private reality will now be completely separate.

There's only one thing that makes me think this won't happen. I told my vision to a hill farmer I know. "Shucks," he said. "You think I could get some of those logs when the fellow's through with them? My furnace eats wood something awful."

[1964]

Second Person Rural

Best Little Woods Tool Going

🐾

I LEARNED A LOT of woodcraft when I was a boy. Between the Boy Scouts, my father (a fine woodsman condemned to spend most of his life at a desk in New York City), the books of Ernest Thompson Seton, and the L. L. Bean catalogues with which our house was constantly awash, I knew my way around for a suburban kid. I could make a fire on a wet day, find east by looking at the top twigs of young hemlocks, space my chopping cuts accurately on a birch log twelve inches in diameter, and so on.

Among other things, I knew a fair amount about woods tools. This was the pre-chainsaw age (at least among Eastern sportsmen: I guess a few loggers were already staggering around with the early two-man models), but axes, hatchets, saws, and wedges were familiar objects almost from birth. By the time I was twelve, I could make a straight cut with a one-man crosscut, or pull lightly and smoothly on one end of a two-man saw – especially when my father was on the other end. I could place the splitting wedges with moderate accuracy in a knotty beech log. I had views on the weight of axheads. Give me a sharp bucksaw, and I would make you a fast pile of stovewood.

I had never heard of peaveys.

The peavey is an instrument worth hearing of. I won't say they're indispensable for the casual tree cutter – but they do make work in the woods a lot handier. The bigger the trees you fell, the handier a peavey is. Anyone who gets firewood from trees much over a foot in diameter could use one. Anyone who cuts even an occasional sawlog could use two.

Archimedes once boasted that if he had a long enough lever, he could move the world. Archimedes would have loved peaveys. A peavey can be defined as a lever with a built-in fulcrum. It consists of a heavy wooden handle three to four feet long, with a steel head. Mounted on one side of the head is a steel hook that will bite into a log. In this form it has existed for a very long time – lumbermen in classical Greece may have used bronze-headed ones, for all I know – and this primitive version is called a cant hook or cant dog. Then in 1858 a Maine blacksmith named Joseph Peavey got the idea of turning the fixed hook into a swinging hook that will bite easily into a log of any size. Eureka! The peavey.

I was around thirty when I saw my first peavey in action, and I wasn't very impressed. What I saw was a Connecticut farmer using one in its commonest and humblest function, to roll a log over. I had an Abercrombie and Fitch reaction. Or perhaps the same reaction a cross-country skier has when he sees his first snowmobile. The damned things just weren't sporting. I knew how to roll a log over; my father had taught me. What you do is to leave the stub of one branch when you're limbing – say, a stub three to four feet long – and then you just roll the log over with that. Organic log handling, so to speak.

A few years later I went out and bought a peavey. (You can get one for as little as fifteen or twenty dollars; the best big ones from Snow and Nealley, Bangor, Maine, might cost thirty dollars.) These days I own two. What happened in the interim was that I made the definite switch from crosscuts and axes to the chainsaw, and I had gradually begun to cut much larger trees than I ever did before.

The particular large tree that led me to buy my first peavey was a white ash growing in a fence line between two fields. Two and a half feet in diameter, and probably sixty feet tall. I wanted it out of there, partly because a couple of nice ten-inch sugar maples were growing practically under it, and I wanted them to have the light. But mostly because I needed some firewood in a hurry, and ash doesn't need much drying. (Though, as I discovered, it will burn a lot better if you *do* dry it.)

In my old crosscut days, dealing with that tree would have been a half-a-week project. An ash that size has two cords of wood in it, and weighs three tons. In fact, I might have just looked at it respectfully, and found something smaller. But by now I was fairly competent with a chainsaw. I strolled out and started cutting.

In a couple of hours I had the tree down and the whole top converted into firewood. Then it was time to attack the trunk. Since there were no limbs for the first thirty feet, I didn't have much limbing to do. In ten minutes I had the trunk bare (except, of course, for the turning stub I had left), and in another twenty I had all my cuts made partway through the trunk. Now to turn.

I gave the stub a firm push. Nothing at all happened. After a while I got my wife, and we both pushed. We couldn't even rock it much.

I have a standard procedure I use in such contretemps. It may lack macho prestige, but it works. I have three friends in town — two native, one not — who are clever at solving problems, and what I do is I call one of them. This particular day I called Tom Pinder, because I happened to know he was home putting a new roof on his house. (I did have the grace to wait until I knew he'd be in the house having lunch.) He came right over, bearing his peavey. He set the hook confidently, just about halfway up the trunk, and pushed. It *was* a big tree, and he did have to strain a bit, but it rolled obediently over. The next day I owned a peavey myself.

Loggers, as I understand, routinely use their peaveys to do everything but pick their teeth. Or at least they used to. There was even supposed to be one cook on the old Penobscot drives who stirred the big kettle of baked beans with a peavey. For the first few months I used mine only to roll logs. Then one day I made a mistake cutting a fifteen-inch red maple and got it hung up in a neighboring oak. I've lodged a good many trees, one time and another, and I am very familiar with the dangerous and humiliating business of cutting short sections off the bottom of the lodged tree and then hopping aside as the now unsupported top thumps down a few feet. I've even pushed a few very small lodged trees over by brute force.

But this one was far too big for me to push over, and it was still so nearly vertical that the idea of cutting a piece off the butt made me nervous. I've seen a man on whom a tree fell.

Once in a while I get an idea of my own. I got one now. Holding my peavey horizontally, I set the hook into the leaning tree about three feet up, and rotated it. The maple majestically spun like a giant top, came clear of the oak, and crashed on down right where I meant to have it in the first place. Thousands of loggers, I now know, have been doing the same for the last 122 years – but for me it was a true eureka moment.

There's a third thing I have learned to do with peaveys that is even more gratifying than the first two. This is to use them in low-technology log loading.

Maybe once a year I get an urge to cut a few white pines or wild black-cherry trees and have them sawed into lumber. Since Gary Ulman's sawmill is only two miles away, it's an urge I can gratify. But no one with a logging truck is going to come pick up my miserable four or five logs – or, at least, if someone did, it would cost so much that my boards would wind up more expensive than first-quality four-sides-planed stuff at the most expensive lumberyard in the Boston suburbs. Instead I take them over in my pickup.

But how do you get twelve-foot sawlogs off the ground and into the back of a pickup? There are various ways. Enough helium balloons will do it, or a forklift truck, or any of the larger-model helicopters. On one memorable occasion, I had six Dartmouth freshmen out for a picnic and got their help. Seven people can lift quite a heavy log, hard though it is for human arms to get a grip. A much simpler method, though, is to get one friend and equip him or her with your spare peavey. Then you back the pickup up to the smaller end of the log. One of you stands on either side of the log, and you hook on with your peaveys. You can get a splendid grip with the peavey, and the two of you can load it with ease. You have to expect a few grins, of course, when you trundle in to a sawmill with four small logs on a pickup – but I *have* known a real logger who was there with his monster truck and fifty big logs and cherry picker to get so involved helping me unload that he almost walked off with my spare peavey. I think he was having a fit of nostalgia.

Not that that's the only way to load logs with a peavey. That's just the way you do it when you've got a pickup and have to slide them in through the tailgate. With a more versatile truck, namely one that has removable sides, you can roll them on.

I did my first roll-on loading just about a year ago. I had gotten carried away and dropped a couple of white pines that even an Oregon logger wouldn't have sneered at. I don't say he'd have been *impressed*, but he wouldn't have called them Vermont toothpicks, either. They were nice tall pines, each with three sawlogs. The butt log of the bigger one was just about twenty inches in diameter. There's no way two people are going to pick up one end of a log like that and hoost [*sic*] it into the back of a pickup.

Fortunately, my neighbor George deNagy has an old one-ton farm truck, the kind that looks much bigger than it is because the bed sits up over the wheels. Except that it gets five miles to the gallon and won't start in cold weather, it's a lovely vehicle. And, of course, you can take the sides off.

George loves challenges. He was glad to come rumbling into my woodlot (that's no cliché – farm trucks invariably rumble; it's because of those detachable sides) and pull up next to my modest log pile. We then cut three ten-foot hemlock saplings. Two are all you really need; the third is just a little piece of insurance in case you've cut the saplings too small and one of them happens to break just as you're rolling a log up.

We laid the saplings against the side of the truck so as to form a ramp. One went at each end of the truck bed, and the insurance sapling rested in the middle. Then we hooked on to the first and biggest log with our peaveys and began to roll. It worked beautifully. The bark on a green sapling makes a nice corrugated surface, and the log never even threatened to slip. (I'm told that people who use sawed timbers for the ramp sometimes find the log skidding right back down onto their feet.) In twenty minutes we had loaded all six logs and were on our way to Gary's mill. Two weeks later I had as handsome a pile of clear pine lumber as a man could ask for. Score another victory for the peavey.

I have probably now reached my personal limit of peavey skills. After all, I started late. But there is one other of their many uses that I hope to witness sometime in my life. Peaveys are the tool of choice in river drives. I once heard a lumberjack's song about breaking a logjam. The boss wanted to dynamite it – which of course would have damaged some of the logs, besides being crude and noisy – and the lumberjacks said no, they could pick it apart. Guess with what. As the song put it:

> But before you try the powder,
>   Or to break her with the juice,
> Hand some peaveys to the river rats and jacks.
> They will roll her and they'll crowd her
>   And they'll break the timber loose;
> Yes, they'll break her, or half a hundred backs.

As poetry that's on the weak side, but as a vision it's something else. Fifty men out on a river with peaveys, swarming over a million board feet of tangled pine logs would be a sight worth seeing. If gasoline goes enough higher, I might even get to.

[1979]

Maple Recipes for Simpletons

∽❦∽

THERE ARE A lot of maple recipes in existence. Someone once gave me a book that contains at least three hundred – in fact, that's all the book *does* contain. There are recipes for Maple-Cheese Spoon Dessert, and for Modern Maple-Pumpkin Pie. For baked squash covered with crushed pineapple and doused with maple syrup. For peanut-butter cookies. Even directions for a truly revolting salad dressing. (You mix cream and lemon juice, and then add a big slug of maple syrup. Oil and vinegar with a discreet touch of garlic is more my idea of a salad dressing.)

I yield to no one in my admiration for maple syrup. I've been making it for fifteen years; and even with my little rig, total production now comes to many hundred gallons. I have gradually learned to make not only syrup, but tub sugar, maple candy and finally, just in the last few years, the highest art of all: granulated maple sugar that pours as readily as the white stuff you get in a five-pound bag at the store. These products taste, if anything, even better as the years go by.

But all the same, I view most maple recipes with dark suspicion. Too many of them put a noble product to unworthy, not to say peculiar, uses. Many also ignore the fact that maple syrup currently costs about eighteen dollars a gallon, and is thus a pretty expensive sweetening agent.

Take those peanut-butter cookies. To make one batch requires half a pint of maple syrup, and all you wind up with is something

41

that tastes like sweet peanut butter. Ten cents' worth of cane sugar could handle that job – and it's just the sort of humble task cane sugar was born for. As for mixing syrup with crushed pineapple and plastering it on hunks of squash, I'd as soon mix twelve-year-old Scotch with diet Pepsi.

I certainly don't claim all maple recipes are like that. A good maple cake, maybe with some butternuts in the frosting, is one of the joys of life. And I've had a maple charlotte I would walk several miles to have again. These are splendid uses of syrup; the maple flavor comes out, if anything, enhanced. My only problem is that I am not personally competent to make either a cake or a charlotte.

However, there are some recipes that I *can* handle and that are maple-enhancing. I propose to share three of them. Two are my own discoveries, the third is a standard rural treat. All three are notably easy to prepare. In fact, they are so simple that any cook is going to regard the word *recipe* as absurdly out of place. So perhaps I should instead say, here are three good uses for maple syrup.

The first recipe is for Vermont baklava. Greek baklava (which came first by about five hundred years) is a many-layered pastry soaked in and fairly oozing honey. Vermont baklava is less complex. The ingredients are a loaf of good-quality white bread (home-made, Pepperidge Farm, Arnold, etc.) and a can of maple syrup. To prepare it, you take two slices of bread from the loaf and place them in your toaster. Set the toaster on medium. When the toast pops up, remove and place on a plate. Then cover each piece generously with maple syrup. Wait two to three minutes for it to soak in. The baklava is now ready for consumption.

Two important tips: On no account heat the syrup, and on no account butter the toast. It is essential that the only ingredients be white bread and room-temperature syrup.

I stumbled on the recipe for Vermont baklava about ten years ago, when I first became a commercial-syrup producer. People began to stop by my farm to buy syrup. I would ask them what grade they wanted; and naturally enough a fair number didn't even know there *were* grades. My usual impulse was to give them a sample of each grade. But straight syrup from a spoon is a little overwhelming, and I certainly wasn't going to fire up the stove and make a batch of pancakes for every visitor. One day it occurred to me to try toast. I omitted butter simply because we happened to be out. And I then discovered that toasted white bread is one of the great vehicles for maple syrup. One gets the full brilliance of the flavor – if you'll forgive the arty term – and one gets something else that I have never experienced elsewhere except in tasting partly finished syrup in an evaporator. Poured at room temperature over toast, maple syrup by itself seems to have the qualities of butter, along with its own characteristics. The dish thus recommends itself especially to those who love butter but avoid it on account of their fear of polysaturated fats.

May be made with Fancy, A, or B. Not suggested with Grade C.

The second recipe will be of interest only to those who like to eat sliced bananas and milk. And even within that already limited group, only to those who feel that sliced bananas and milk go much better with brown sugar than with white sugar.

I have felt thus since roughly the age of six. In those early years I was likely to have a base of cornflakes under the bananas and brown sugar; since about sophomore year of college I have omitted the cornflakes. They only get soggy, anyway.

The ingredients called for in the second recipe are one or more ripe bananas, a supply of whole milk, and some dark maple syrup. Slice the bananas in the usual way, add the normal quantity of milk, and then pour in a couple of tablespoons of maple syrup. (Right in the milk? You feel it might be like adding syrup to crushed pineapples and squash? I assure you it is not.)

The recipe is again one I stumbled on. One night a couple of years ago I happened to be fixing dinner alone. My wife and daughters had gone to a fair, and wouldn't be back until late. I usually figure on a maximum preparation time of ten minutes when I'm fixing dinner alone, so as to waste as little time as possible indoors. The menu this particular evening was a hamburger, to be followed by sliced bananas, milk, and brown sugar. Then I couldn't find the brown sugar. Not only no real brown sugar, but not even any of that light-tan stuff that will do in a pinch.

I already had the bananas sliced. Some kind of sweetening was necessary for the milk. We happened to have an open jar of Grade C in the pantry. I went and got it. At first bite I realized that for over forty years I had been having second-class bananas and milk. With brown sugar it's good. With dark maple syrup it's better.

May be made with Grade C or Grade B. Not recommended with Fancy or A.

The third recipe is the New England equivalent of sweet-and-sour pork in a Chinese restaurant, and it is a traditional spring dish. Warning: Anyone on a diet – in fact anyone who is not something of a glutton – should not even read about this dish.

Ingredients: a dozen plain raised doughnuts (two dozen, if more than four people will be present), a large jar of dill pickles, a quart or more of maple syrup.

First you boil the syrup down by about one-third, so that it has the consistency of a sugar glaze. Meanwhile, quarter the dill pickles and put them in a dish in the middle of the table, right next to the unsweetened raised doughnuts. Then, while it is still warm, you put some syrup in the bottom of a soup dish for each person.

Everybody then takes a doughnut, dips it in his or her bowl of syrup, and begins to gorge. After every two or three bites – or at a minimum twice per doughnut – you stop and eat a bite of pickle. With this constant resharpening of the palate, it is possible to eat an astonishingly large number of doughnuts. Stop just before you are comatose, and conclude with a cup of brewed coffee. Then retire to bed.

Should be prepared with Fancy or A. B will do, though not as well. C is not recommended.

This by no means exhausts the list of simple maple recipes – a small quantity of B or C does wonders in a pot of baked beans, a little A on popcorn beats Cracker Jack hollow. But it's enough to use as much spare syrup as most people are going to have in the course of a year.

[1980]

Vermont Silences

THERE IS AN OLD story about two Vermont farmers who lived a mile apart – one west of the village, and the other east of it. Since rural free delivery didn't exist yet, each had to come into town to get his mail. Every weekday for twenty years Eben would finish morning milking and come striding into the village from the west, while Alfred did the same thing from the east. Since both were punctual men, they invariably met in front of the post office at nine A.M., just as the last letters were being put up. They'd say good morning, go in and get their mail, and stride off home – one west and one east.

One morning during the twenty-first year, however, Eben came stumping out of the post office and, ignoring his usual route, started briskly south, down the state highway. Alfred stared after him for a second, and then called, "Eben, where on earth ye going?"

Eben whirled around. "None of your goddamned business," he snapped. Then he added, visibly softening, "And I wouldn't tell ye that much if ye wan't an old friend."

This story conforms perfectly to the stereotype of conversational habits in the country. City people talk a lot, the belief goes, but rubes are closed-mouthed. They think they've had a big conversation if one person says he hears Harley's brown cow is going to calve, and the other answers, "Ayuh."

For the first year or two that I lived in the country, I believed firmly in the stereotype. (Even now I wouldn't call it wholly false.

45

There *is* less badinage in the average milking parlor than in the average cocktail lounge.) But I have since come to realize that words can fly as fast in the country as in town. It's just that rural conversation operates under a rather peculiar set of rules. And the rules do impose certain silences.

The first rule is that only the person who's supposed to be talking does. The others keep quiet. For example, say a fellow who might be going to mow your hay comes by on a Saturday afternoon to discuss the terms. His wife and her brother are in the car; they're on their way to go shopping.

In the suburbs where I grew up, there would be some kind of general introduction. "This is my wife, Alice, and my brother-in-law, Fred." "Pleased to meet you, Alice. Hello, Fred."

There is none of that here. The wife and the brother-in-law sit out in the car, not saying a word. You never even learn their names. Weird, silent, unsocial country people, you think. But it's not that at all. It's just that this is not their deal. If it were, they'd have plenty to say.

I first realized that when I lived for a couple of years in a house that belonged to a rural utility company. This hick utility owned two small reservoirs, and provided water for a New Hampshire town of about seven thousand. The house was right by the lower dam, and normally it was occupied by one of the three employees of the company.

But the guy who'd been living there had just bought a trailer, and the other two weren't interested, so the company rented it to me. It was a terrific deal. I paid a hundred dollars a month rent – this was 1961 – and meanwhile the company paid *me* twenty-five a month to keep people from swimming or fishing in the reservoir. Naturally I interpreted "people" to mean the general public, but not me, my family, or a few close friends. So for seventy-five dollars a month I had a nice little house plus a private ten-acre lake. Good clean water to swim in, and really great bass fishing.

Every morning the foreman, a man named Asa, would come by to check the dam and the chlorination unit. He always came in a pickup truck, and he always had his second-in-command with him. (It took two people to deal with the dam gates.) If I happened to be outside when they came, Asa and I would usually have a little chat. I'd tell him how many trespassers I'd driven away; he'd tell me stories about New Hampshire years ago, or funny things that happened while they were fixing pipes in town. The second-in-command never spoke. I didn't even know his name. One of those weird, silent country people.

Then after about six months, Asa was sick one day. But the truck came as usual. The second-in-command was driving, and he had the third employee of the company with him. I was outside when he arrived. As he drove into the barnyard, he stuck his head out the truck window and called cheerily, "Nice mornin', ain't it? Looks like we'll get spring, after all." Ten minutes later I knew quite a lot about Denny. We'd swapped a couple of jokes, traded views on baled versus loose hay. He wound up inviting me to come take a look at his farm sometime.

As for the third-in-command, a kid about twenty, he hadn't said a word. I didn't even know his name.

Another rule is connected with rural stoicism. This rule says never admit to caring too much about anything, because if you do, you'll probably lose it. Hence you simply never encounter the phenomenon I used to in the suburbs, where sometimes at a party one of the guests would suddenly look up and say, "Isn't this *fun!* Aren't we all having a marvelous time!" Except once from a summer person, I have never heard such a remark at a church supper, or at the square dance we have at the end of Old Home Day each summer, or at any Vermont festivity whatsoever. I have seen a good many shining eyes, but heard no gush.

The working of this rule is perhaps clearest among country children. If you invite a city kid to go to the circus, that kid may perfectly well whoop with joy, ask if he can bring a friend, turn cartwheels, if young enough even kiss you. If you invite a country kid, at lease in Vermont, the answer is much lower key. "I don't care," he or she says.

47

What that *means* is "Yes, please, I'd love to go, and you're wonderful to ask me," but he's not going to spill all that out. You have to get the message from his eyes. You have to look pretty carefully, too, because if he didn't want to go, he might well say the same thing. Only then it would mean something like "I'm getting a little old for the circus, but if you're determined to take me, I guess I can stand it."

This trick of speech makes it easy to sort native from non-native children. Just invite all the kids in town to a soda fountain, and ask who wants an ice-cream cone. All the ones who shout "Me" or "I want choc'late!" or who groan and say it's too soon after lunch are the children of immigrants from New Jersey. All the ones who say "I don't care" are natives.

I still haven't mentioned the most important rule. This is the one that's descended from the work ethic. It says that conversation should never be sought for itself, but should just sort of happen. Deliberately to plan some occasion when you do nothing but talk (e.g., a cocktail party) is certainly foolish and probably immoral.

The best way to let it happen is to share a job with someone. *That* you can plan – a quilting bee or a barn raising is just fine – but officially you're there to work on the quilt, or nail rafters; and if there happens to be a steady stream of conversation while you stitch or pound, well, you're just as surprised as everybody else. You came to work.

Hence all my conversations with Asa were just before or just after he checked the dam. When I talk with Rodney Palmer, it's at his garage, and he generally has a wrench in hand. The best conversation I ever had with Wesley LaBombard was while we were planting trees. It lasted seven hours, and ranged from the existence of God to how to raise chickens. Meanwhile, we planted four hundred red pines.

The most remarkable example I know, though, of work freeing a countryman to talk concerns a master mason whom I shall not name. He's a man in his sixties, taciturn even by the standards of rural Vermont. For forty years he has been a silent attender of church suppers (when he goes at all), a silent presence in some summer person's yard, rebuilding a stone wall.

But a couple of years ago a friend of mine who's pretty handy himself got this man to teach him how to build chimneys. In effect, he apprenticed himself. Well! The first couple of days they worked mostly in silence, except for businesslike remarks about hearth laying and lapped courses. Somewhere on the third day they began to trade views about coon hunting. By the end of the week they were into politics.

It was the middle of the second week that the old mason really opened up – never, of course, ceasing to work. At the time he began, they were cutting tiles with a masonry saw, and you are to imagine a noise somewhere between a dentist's drill and a jet takeoff in the background.

It turned out the old man had worked for a couple of years in Boston when he was just out of high school. He had proved very attractive to Boston girls, but in his stiff country way he had repelled their advances. (This, of course, had made him even more attractive to certain ones, who liked challenges.) For forty years he had been wondering whether he had done the right thing to repel those girls, or whether he had been a young idiot. He could remember some of the evenings more or less minute by minute, and he reported every detail. Still, of course, using the masonry saw.

My friend says he talked a blue streak. He would describe a date with one of the Boston girls – it was like hearing a documentary, my friend says – but at the crucial moments he was fairly often sawing a tile. So that what my friend heard went something like this:

"It was about midnight and her folks was all asleep. We were standin' in the hall. I was just going to tell her goodnight, and she says to me" – here he started cutting a tile, and Tom didn't catch another word until the tile fell apart. "Well I didn't quite know what to make of that, so I says t' her, 'Mary, I don't see as you have any reason to'" – *neeeeeeyowhh* – *neeeeeeyowhh* – "and b'gosh if she wasn't trying to drag me" – *neeeeeeyowhh* – "so, Tom, what would you 'a done if you'd been me?"

SECOND PERSON RURAL

Doesn't that sound almost as lively as the average conversation in a bar? Maybe even livelier than some? You know it does. If you want *real* talk, forget the city. Move to the country, and get yourself a job on the road crew (you'll make about $3.25 an hour) or helping some old-timer sharpen axes. During the intervals, when you can hear, you'll learn what country conversation is really like.

[1979]

The Rural Immigration Law

EACH MAN KILLS the thing he loves, Oscar Wilde wrote in a poem that later became a popular song. As a general statement, this won't do. Lincoln didn't kill the Union; lots of men don't kill their wives; so far from killing the ERA, Betty Friedan and Kate Millett have worked hard to keep it alive.

But practically all tourists and most people who move to the country do kill the thing they love. They don't mean to – they may not even realize they have done it – but they still kill it.

The tourist does it simply by being a tourist. What he loves is foreignness, difference, the exotic. So he goes in search of it – and, of course, brings himself along. The next thing you know there's a Holiday Inn in Munich.

The case with people who move to the country is more complicated. What they bring along is a series of unconscious assumptions. It might be better for rural America if they brought a few sticks of dynamite, or a can of arsenic.

Take a typical example. Mr. and Mrs. Nice are Bostonians. They live a couple of miles off Route 128 in a four-bedroom house. He's a partner in an ad agency; she has considerable talent as an artist. For some years they've had a second home in northern New Hampshire. The kids love it up there in Grafton County.

For some years, too, both Nices have been feeling they'd like to simplify their lives. They look with increasing envy on their New Hampshire neighbors, who never face a morning traffic jam, or an

51

evening one, either; who don't have a long drive to the country on Friday night and a long drive back on Sunday; who aren't cramped into a suburban lot; who live in harmony with the natural rhythm of the year; who think the rat race is probably some kind of minor event at a county fair.

One Thursday evening Don Nice says to Sue that he's been talking to the other partners, and they've agreed there's no reason he can't do some of his work at home. If he's in the office Wednesday and Thursday every week, why the rest of the time he can stay in touch by telephone. Sue, who has been trapped all year as a Brownie Scout leader and who has recently had the aerial snapped off her car in Boston, is delighted. She reflects happily that in their little mountain village you don't even need to lock your house, and there *is* no Brownie troop. "You're wonderful," she tells Don.

So the move occurs. In most ways Don and Sue are very happy. They raise practically all their own vegetables the first year; Sue takes up cross-country skiing; Don personally splits some of the wood they burn in their new wood stove.

But there are some problems. The first one Sue is conscious of is the school. It's just not very good. It's clear to Sue almost immediately that the town desperately needs a new school building – and also modern playground equipment, new school buses, more and better art instruction at the high school, a different principal. Don is as upset as Sue when they discover that only about 40 percent of the kids who graduate from that high school go on to any form of college. The rest do native things, like becoming farmers and mechanics, and joining the Air Force. An appalling number of the girls marry within twelve months after graduation. How are Jeanie and Don, Jr., going to get into good colleges from this school?

Pretty soon Sue and Don join an informal group of newcomers in town who are working to upgrade education. All they want for starters is the new building (2.8 million dollars) and a majority of their kind on the school board.

As for Don, though he really enjoys splitting the wood – in fact, next year he's planning to get a chainsaw and start cutting a few trees of his own – he also does like to play golf. There's no course within twenty miles. Some of the nice people he's met in the education lobby feel just as he does. They begin to discuss the possibility of a nine-hole course. The old native who owns the land they have in mind seems to be keeping only four or five cows on it, anyway. Besides, taxes are going up and the old fellow is going to have to sell, sooner or later. (Which is too bad, of course. Don and Sue both admire the local farmers, and they're sincerely sorry whenever one has to quit.)

Over the next several years, Don and Sue get more and more adjusted to rural living – and they also gradually discover more things that need changing. For example, the area needs a good French restaurant. And it needs a *much* better airport. At present there are only two flights a day to Boston, and because of the lack of sophisticated equipment, even they are quite often canceled. If Don wants to be really sure of getting down for an important meeting, he has to drive. Sue would be glad of more organized activities for the kids. There's even talk of starting a Brownie troop.

In short, if enough upper-middle-class people move to a rural town, they are naturally going to turn it into a suburb of the nearest city. For one generation it will be a very nice and a very rustic suburb, with real farms dotted around it and real natives speaking their minds at town meeting. Then as the local people are gradually taxed out of existence (or at least out of town), one more piece of rural America has died.

This is happening to large parts of New England at the moment. The solution, as I see it is a good, tough immigration law. It wouldn't actually keep Don and Sue out, it would just require them to learn rural values before they were allowed to stay. When they moved to the country, they would be issued visas good for one year. At the end of that year, they would have to appear before a local board composed entirely of native farmers, loggers, and road-crew men. They would then present evidence of having acclimated. For example, they could show proof of having taken complete care of two farm animals of at least pig size, or of one cow, for at least nine months. Complete care would be rigorously interpreted. Even one weekend of paying someone to feed the pigs or milk the cow would disqualify them. (An occasional trade, on the other hand, would be acceptable. Don and Sue could take care of a neighbor's stock one weekend, and thus earn the right to be away the next, while he looked after theirs.)

Such a rule might work a hardship on elderly people moving to the country – say, if Sue's parents decided to come up from Baltimore. For them there would be an appropriate modification. The old couple wouldn't have to learn how to handle cattle at sixty-eight and sixty-three. They wouldn't even have to get up on the roof of their house, like natives, and shovel snow or replace missing shingles. But they would have to do undelegated work. For example, if they both worked in the kitchen at all church suppers during the first year, personally cooking beans and making the red-flannel hash, that might earn them a visa renewal. A cash donation would get them nowhere.

What if the board didn't pass you? I'm kindhearted. I wouldn't say you had to clear out immediately. It's just that your taxes would automatically double. They'd stay double until you passed your preliminary test.

And if you did pass? Why then you'd get a five-year visa under the same conditions. By then a seasoned pig raiser, woodlot manager, or church-supper worker, you would appear before the board a second time. If the board approved, you would then be an Accepted Resident – and, incidentally, perfectly free to spend all your time playing golf, trying to turn rural schools into suburban schools, etc., etc. But I'm guessing that very few immigrants would. It's so much more interesting to keep pigs.

What about all the second-home owners who aren't residents anyway? That's easy. Double all their taxes right now.

[1980]

54

The Year We Really Heated with Wood

❦

To boast in 1978 that you have a wood stove is about like telling people proudly that you own a TV set, or that your kitchen has a sink. By the latest estimates, more than half of all rural New Englanders run a wood stove these days. In places such as northern Michigan the proportion may be even higher.

But to *have* a wood stove and to depend on wood for your principal heat are two very different things, as my family and I discovered last winter.

We are old hands with wood stoves. For the last fifteen years we've been living on a Vermont farm. The house had an oil furnace when we bought it – and the farm had a big woodlot. No oil wells, however. So fourteen and a half years ago we set up a couple of stoves: one in the kitchen and one in the living room. We've used them, too. Saved a significant fraction of our heating bill, and all that. Never had any problems. A few years ago we even added a third stove – a nice little Jøtul in the guest room upstairs, which has always been cold. But we ran it only when we had guests.

Then came the well-known rise in the price of fuel oil. By the spring of 1977 the stuff was selling around here for fifty-one cents a gallon, which at the time seemed high. What had been half a game for fourteen years suddenly became serious business.

By then I was an experienced woodcutter. Experienced enough

to know that my woodlot could yield ten times what I had been cutting, and actually benefit from the process. Why not step up production? So in the spring of 1977 we decided to switch from System A, an oil furnace supplemented by wood stoves, to System B, wood stoves supplemented by an oil furnace. This is the story of what happened. Or part of the story, anyway. I haven't been able to get in everything, like my wife's learning to play the piano twenty minutes at a time, and then dashing into the kitchen to wash a few dishes, because none of our stoves really gets to the room where she keeps her piano. But the main elements are here.

March 7, 1977. Today we paid 526 dollars for a new stove. If we're really going to heat with wood, we need a really powerful stove. It's got to handle four rooms downstairs and four more upstairs – everything, in fact, but the kitchen and the guest bedroom. So we've taken out our hundred-year-old cast-iron parlor stove and replaced it with a brand-new Defiant. As a fringe benefit, we will gain some space. One of the Defiant's advantages is that you can mount a heat shield behind it, and then put it practically up against a wall.

March 8. I took the day off from work and spent it entirely on the stove. If it's to go where my wife wants it, we need a new stove hole on the left side of the chimney. Our neighbor Lee Ilsley came and cut a perfect round seven-inch hole through the chimney bricks with hammer and chisel. Then I helped him install the thimble. The rest of the day I devoted to buying stovepipe, setting it up, building a new and larger woodbox, and so on. Tonight it all looks worth it. The Defiant is tucked away beside the chimney, looking very handsome. The back is only five inches from the brick house wall.

The old stove used up about a third of the room.

April 8. Maple-sugaring ended two days ago. Starting today, I am spending my free time cutting next winter's wood. I already have about four cords I cut last fall – but with System B I figure I will need at least twice that much. It is already late to be cutting wood that will be dry enough.

56

The Year We Really Heated with Wood

April 9. Today is Saturday, and I spent the whole day in the woods. It would make sense to cut all the trees I'm going to, right now, so they could start drying. But psychologically that doesn't work for me. It's too much like an assembly line. I prefer to take one tree at a time, fell it, buck it up, and do all the splitting, before I move on to another. Today I did about a tree and a half. Both were red maples eighteen inches in diameter. A tree this size yields something over two-thirds of a cord.

April 23. I'm past the eight-cord mark, and still cutting. My wife has started coming out with me, to help with the splitting. She learns fast.

May 2. We've decided to cheat a little on the wood. Counting last fall's, we have nearly ten cords of long wood for the Defiant cut, split, and stacked. That was fun.

But I'm getting tired of cutting very short pieces for the two little stoves: the Ulefos in the kitchen and the Jøtul in the upstairs wing. So we have ordered a truckload of scrap from Malmquist's mill, four miles away. It's what is left over after they make blanks for chair legs, and it's mostly pieces of rock maple and yellow birch six to twelve inches long. You get about a cord and a third for forty dollars.

May 28. We've had a dry spell, when I could get my truck into the woods. Most of next winter's firewood is now stored in the barn, ready for use.

It's been a problem spreading all that weight around. Our barn is built right onto the house, and it has a cellar, just like the house. Earlier owners kept their cattle down there. I don't want to keep our wood down there, because it would mean lugging every arm-load up the stone cattle ramp next winter. But I'm scared to keep too much in any one place on the main barn floor, for fear of breaking the beams. Result: I have five different piles of about half a cord each, plus two cords stacked on the ramp itself. Another cord and a half is outside, stacked against the west wall of the barn.

There's a connecting shed between the house and the barn (the cellar goes under that, too), and that's where we used to keep practically all of our wood. This year it just has the short wood for the two little stoves. About three cords of it, counting the load we bought. The maple from Malmquist's is awfully heavy. I worry a little about having stacked it six feet high.

May 31. We haven't lit a fire even once this week. Heating season is over. I cleaned the stoves, and spread the ashes in the back pasture. Lilacs are at their height, and apple blossoms almost over. We've got peas up six inches.

September 12. I haven't thought much about heating systems for the last three and a half months. Been too busy gardening and raising sheep. Last night's frost and this morning's kitchen fire reminded me, though. It's time for the next project, which is to insulate the cellar. People have warned me that if we don't run the furnace regularly, our cellar will freeze. But I have a solution. Most of the cold gets in where the house cellar meets the barn cellar. I am going to build an extra wall eighteen inches outside the existing wall, and fill the space with sawdust. The old-timers used to keep ice all through the summer by burying it in sawdust; I reckon I can keep icy air out the same way.

September 17. The wall is built. It took a full pickup-load of sawdust to fill the space – and it must be one of the least expensive insulating jobs on record. A hundred and twenty cubic feet of good hardwood sawdust cost me six dollars.

I may be imagining it, but the shed floor seems to have developed a slight slant.

October 1. Cold, steady rain for the last two days. We've had two stoves going, and the house is beautifully warm. Who needs a furnace?

October 4. The rain has continued, and we are using wood at a good clip. It is already clear I don't have enough stored. No problem. I've got plenty more cut and stacked in the woods. All I need is a few dry days to take my truck out and get it.

October 5. Perfect sunny day after a sharp frost. I went out at dawn, while the ground was still hard, and brought in two more cords. It's stacked on the front porch, leaving just enough space for the door to open freely.

I am not imagining the slant in the shed floor.

October 6. After work, I went down to the barn cellar with a flashlight. All three beams under the shed are badly bent. The central one has cracked, and is on the verge of breaking. I borrowed two jack posts from my neighbor Barbara Duncan, and spent the evening trying to jack the center beam back up. No luck. The woodpile up above is just too heavy. But at least I don't think the floor will sag any more. I'll worry in the spring about how to straighten the beams.

November 24. Snow today. We just smile and stoke our stoves. We have yet to use the furnace this winter.

December 8. I didn't use screws when I mounted the Defiant's stovepipe, and a couple of sections are beginning to come apart. We are not eager to have the house burn down. Annemarie let the fire go out this morning, so that this evening I could take the pipe down and reassemble it with screws.

Since it was five below zero when she got up, this naturally meant starting the furnace. I wasn't altogether sorry – even though I had hoped to go until Christmas. Our water pipes run under the kitchen, just inside the new sawdust wall. Somehow the cold is getting through. Last week I took a thermometer down there. This morning it registered thirty-six degrees. A little hot air in the furnace pipes will be a good thing.

December 10. We've been back on stoves for two days, and it is now thirty-three degrees inside the sawdust wall. I consulted my friend Tom Pinder, who is clever about such things. With his help, I've vented the clothes drier directly into the cellar. (More accurately, he did it, while I passed him tools.) Come spring, we'll shift the vent back outside.

December 11. The coldest day so far this winter. Minus twelve. The pipes to one bathroom were frozen. The water pump, miraculously, was not, though the cellar thermometer read twenty-eight degrees this morning. I took down a bucket of coals from the stove.

December 12. Twenty-four hours of running all three stoves at top heat have failed to melt the frozen pipes. The problem is that the warm air from the stoves can't get at the pipes. They're in the walls – right next to the air ducts from the furnace. No one was thinking about heating with wood when this house was plumbed.

December 13. Shut down the Defiant again, and let the furnace run all day. Even though the temperature never rose above six degrees, we succeeded in thawing the pipes out. Starting today, I am leaving the furnace turned on. The plan is to keep the thermostat set low (the markings only go down to fifty-two degrees, but you can actually set the dial at about fifty) – and then count on the furnace to come on for a couple of hours late each night, just enough to keep the cellar warm and the pipes unfrozen.

January 25, 1978. The system is working perfectly. It doesn't even take much oil. We had our second delivery of the winter today. The tank got topped off in September with 28 gallons, and today it took 140 gallons. And we are halfway through the winter! Other winters, we have used five or six hundred gallons by now, even with the old parlor stove going most of the time. (And if we had ever tried a completely stoveless winter, this big brick house would have drunk 900 gallons by now. I know, because once we rented the house for a year and our tenants never lit a match. They got through almost 1,800 gallons in what was a fairly mild winter.)

February 4. The system *was* working perfectly. Yesterday the temperature dropped to fourteen below. Today it's twenty below, and windy. Even though I left the thermostat at fifty-eight instead of the usual fifty last night, we have no water this morning. Two buckets of coals in the cellar before I went to work brought the water back by noon – but I don't want to spend my life providing hibachi service for water pipes. Before next winter I shall either put a little stove down cellar or figure out a way to insulate the rest of the cellar walls, or maybe both. This morning it was twenty-six degrees next to the sawdust wall. And thirty-seven at the other end of the cellar, where I keep the wine. California and France make poor training for a Vermont winter. I hope the wine wasn't too upset.

February 16. Another bitter day. We have now used all the wood on the front porch, all the wood on the stone ramp, and three of the other five piles. There's still some stacked outside along the barn wall, which I never had room to bring in last fall – but it is currently buried under six feet of snow, piled up when the barnyard was plowed.

I see two choices. I can dig it out, or I can go to the woods and cut dead elms. Otherwise, we seem certain to run out of wood before spring.

February 25. I chose to cut dead elms. This was a sunny Saturday, delightful to be out, and I spent nearly all day cutting on a low hillside, and then riding the wood out on a toboggan. Got a two-week supply (at February-March burning rates).

March 15. Today we had our third and final oil delivery: 169.6 gallons. Then we turned the furnace off until next winter. I'm no longer worried about pipes – we're already into mud season, and there wasn't even a frost last night. And I'm not worried about wood, either. The dead elm didn't quite last two weeks – but the powerful March sun has already melted the snow away from the barn wall where I had wood stored outside, and I can get at the top rows easily.

We used 337.6 gallons of oil, at a total cost of 178 dollars, 13 cents. This is lower than our oil bill ten years ago. In the winter of 1967-68, despite help from two stoves, we used 1,317 gallons of oil at 16.6 cents a gallon, and it cost us 218 dollars, 34 cents. Who says there's inflation? The *country* may have inflation, but this farm is enjoying deflation.

It's true that I've got to fix the shed floor, insulate the cellar, think about getting another stove, worry about the piano room, plan better storage facilities in the barn, and fetch home ten cords of wood – but, then, I enjoy doing most of these things. Keeps my weight down, and it's a hell of a lot more interesting than jogging.

61

As for the wood itself, it was cut and split last fall and winter. In fact, while I was at it, I cut twenty cords instead of ten. What we don't use, I shall sell. After deducting expenses in woodcutting, and the cost of whatever oil we buy, next winter I expect the process of heating our house to produce a cash profit. We may just blow it on a trip to Saudi Arabia.

[1978]

Postscript, 1980. We have now been through three winters on System B. Only it's not System B anymore, it's System B-plus. Practice does help.

Oil consumption has continued to decline. The second winter it dropped more than a hundred gallons, down to 204. Last winter, with heroic effort, we pushed it on down to 137. This, I suspect, is about as low as it's going to get – at lease if we want to continue using the plumbing.

As to cost, our private deflation continued for a second year. In 1978-79 our total oil bill was 120 dollars, 49 cents. That is surely the lowest oil bill the house has ever had, even though it got its oil furnace back in the bargain days of 1950. It may even be lower than the coal bills the previous owners were paying in the 1940s. And since I sold 600 dollars' worth of wood, we did get our cash profit.

But last winter our slowly descending use curve met OPEC in round three, and was soundly defeated. For that one solitary delivery of 137 gallons, an unfeeling oil company charged us 139 dollars, 74 cents. One more rise of that magnitude and we will probably shift to System C, a wood furnace. (If we do, we'll put hot-water pipes through it, and so get free hot water six months a year.)

Meanwhile, we have been busy learning new tricks about managing with wood stoves. Most important, we have learned how to keep a fire going pretty well continuously. Back in the old days, I would have to split ten or fifteen boxes of kindling every fall – and the first person up every morning normally had to build a fresh fire in both stoves. Now we use maybe two boxes a winter. Every member of the family knows how to cram the Defiant so full at bedtime that there will still be plenty of coals in the morning. And we all know what proportion of the ashes to take out and just how to rake the coals so that starting the day's fire amounts to no more than putting in fresh wood. (You have to put it in right, of course – and not all at once. They don't call it *building* a fire for nothing.)

Furthermore, we run only one stove now, except on the very coldest days. The little kitchen wood stove is still there, and it's very handy for burning milk cartons and the boxes pizza comes in. Between that stove and the pigs, we don't have much garbage in the winter. But as for keeping the kitchen warm, we do that primarily with a tiny fan. It's four inches square, and it's mounted in the top of the doorway leading into the kitchen. Except when the temperature is ten or twenty below zero, it blows in enough heat from the Defiant to keep the kitchen comfortable. I no longer cut much short wood.

At the cost of shutting off one bathroom from mid-December to early March, we have completely avoided frozen pipes upstairs. The cellar is a little harder; and in really cold weather I am still running the hibachi service. That's another inducement to move to a wood furnace.

But even the cellar doesn't cool off as readily as it used to, because we did indeed do more insulating. In the fall of 1978 I repointed the cellar walls, caulked around the windows, and stuffed fiberglass in wherever any would fit. Then, since there was nothing left to do inside, we started on the outside. Late that fall, copying something we'd heard people used to do a hundred years ago, we banked the house with spruce branches. You make a sort of festoon of them, all around the foundation. It looks pretty, like a giant Christmas wreath laid clear around the house.

But spruce branches turn out to be only so-so insulation. So last winter we tried a different system, employing slightly more modem rural technology. I had a bunch of hay bales that had got rained on (which makes them unappetizing to cows – and less nutritious as well), and I gave the house a necklace of hay. Hay bales do a superior job. Cellar temperatures averaged a couple of degrees higher.

But since I hope not to get any hay rained on this year, and I'm certainly not going to use good cattle hay, worth a dollar-fifty a bale, I have an even more advanced plan for next winter. All the leaves from the ten or so maples in the yard I intend to put into those big green plastic bags you see being carted away in suburbs where it's no longer legal to burn leaves. Only instead of carting mine away, I shall stack them like pillows all around the foundations. A big maple drops a lot of leaves. On the back side of the house, which is also the north, I may have leaf pillows three deep and three high. Right up to the windows, in fact.

People who notice the rapid rate at which our Defiant burns wood sometimes wonder if we won't burn ourselves right out of trees in a few years. The answer is no. Unless I get tired of cutting, we can go on this way forever. Our farm includes just about a hundred acres of woodlot. In a cold climate like Vermont's, an acre of trees will add about half a cord of new growth per year. So our annual production is 50 cords – 900 cords of new wood in the eighteen years we have owned the place.

Some of that growth, of course, is in pines and hemlocks, which we would not use for firewood. Quite a lot is in good, straight yellow birch and ash and sugar maple, trees that it would be criminal to buck up for firewood. But at least a third of it is in trees that ought to come down anyway – red maples that are busy rotting at the core, crooked oaks, black cherry that's getting shaded out and dying. I could cut fifteen or twenty cords a year indefinitely, and just be improving a still-increasing timber stand.

In actual fact, I cut twenty-six cords last year – sold seventeen and kept nine. (Grossed 1,124 dollars, 75 cents; netted plane fare for one to Saudi Arabia.) This year I may cut thirty cords. That still won't be borrowing on the future. It will merely be catching up with the past. Of the 900 cords that have grown while we've been here, about 650 are still out there on the stump. A couple of hundred, at least, are in the form of trees that ought to be thinned out. Even though I never want to have a forest that is wholly practical – with no hollow trees for the raccoons and no three-hundred-year-old low-branched maple stubbornly clinging to life and not growing an inch – I still have a mighty backlog to draw on. In fact, from the point of view of getting this farm in shape, fuel oil at one dollar, two cents a gallon is about the best thing that could have happened.

Third Person Rural

Birth in the Pasture

🍀

My farm (the house excepted) used to be exclusively male. It was sort of like those Greek monasteries on Mount Athos where even the cat is a tom and the chickens are roosters. Only with me it wasn't ideological, and it didn't last as long.

Female animals were excluded from the twenty monasteries on Mount Athos under a constitution promulgated in the year 1045, and still in effect now. The idea was, I think, to be even holier than other monasteries. Female animals tended not to appear on my farm between 1963 and 1974, and there was no idea at all. There was simply a timid owner who didn't want to winter livestock. It was fun to raise lambs and calves, and to see them grazing in the fields all summer. No fun (so I thought) to have to feed and water the creatures all winter. So I bought young stock every spring, and either sold or butchered in the fall.

This naturally meant that I bought males. Females are the important sex biologically – Mount Athos wouldn't last very long without the rest of Greece to supply it with monks, cats, etc. – and farmers like to keep their ewe lambs and heifer calves to raise themselves. They are delighted to sell the surplus nine-tenths of their ram lambs and bull calves.

So always excepting the house, it was just us boys on the place. And an ever-changing group of us, too, all but me.

Then one year I realized I didn't want to keep farming in this stop-and-start fashion. I wanted continuity, familiar faces in the

69

barn, some sense of growth. In short, I wanted a year-round farm, and I wanted it badly enough to be a winter servant to livestock. Rather soon thereafter I owned my first heifer calf. Life has been getting steadily richer ever since.

That calf grew up to be a mostly Hereford cow named Michelle. (The females in the house named her.) When she was one and a half, she met her first bull, an older part-Hereford with no name and no manners. He had a long orange penis, though, and knew just how to use it. There in her own pasture Michelle became pregnant.

Mine is a generation that has wanted to be in delivery rooms. Our fathers took our mothers to hospitals, and there surrendered them to masked obstetricians. They themselves sat meekly in waiting rooms. When all the drama had taken place, behind closed doors, they might be permitted to see the new son or daughter through a glass window. Birth was a mystery to men.

My generation wanted to change that, and did. I was present in the delivery room of a humanistic hospital when my elder daughter was born. I was in the big upstairs bedroom at home, helping Dr. Putnam with the delivery – or possibly just getting in the way – when the younger one arrived. What's certain is that I was able to bring my southern wife a bottle of Dr. Pepper fourteen minutes after Amy was born. I liked that.

Seeing how eager men have been to make their way back into the human birth ceremony, it is not too surprising that some of us also want to assist at the birthings on a farm, or at least be present. Certainly I wanted to be around for the arrival of the first calf to be born on my farm in my time.

I knew the arrival date quite accurately. Michelle and the bull had gotten things started on November sixteenth, just after lunch. The gestation period in cows is about the same as in people, or a shade longer. You figure 285 days. The calf would therefore be due around August twenty-eighth. I say "around," but being an almost mystical believer in animal powers, I really expected her to hit the precise date.

Birth in the Pasture

The first step was to make sure I'd be here. As far as daytime went, there was no problem; in late August I'm nearly always home doing farm work anyway. But evenings are trickier. This is the last moment before summer people leave – at least, those with kids in school. They are feeling nostalgic and social. My wife and I usually get asked out as much during the second half of August – say, three or four times – as in all of September and October put together. It's kind of fun.

But first things first. We agreed that we'd accept no invitations for Delivery Day. And to leave a comfortable margin of safety, none for the day on either side.

The next step was to keep Michelle accessible. Early in August I began taking her a scoop of grain every morning, with the idea of encouraging her to hang around the front of the pasture, where I could watch her. This had its tricky aspects, too, since it's a great big pasture, and there were fourteen other cattle in with her. (One was mine; thirteen belonged to a real farmer.) All fourteen were as fond of grain as she, and at least six were her superiors in herd rank. It was not easy to make the snack an exclusive for her. But I mostly did.

What I wanted was about a two-minute period each morning when Michelle was wholly concentrated on eating, and I could examine her. Hence grain. Cows are so passionately attached to the stuff that they will permit their owners almost any liberty, from putting a rope around their necks to looking under their tails, while they are consuming it. The latter was what I wished to do.

Michelle stayed right on schedule. She had sprung bag way back on the fifteenth. Now her udder got even fuller. On the twenty-sixth of August, two of her teats were stiff with milk. On the morning of the twenty-seventh, all four were. No dilation when I looked under her tail, though. A cinch for the twenty-eighth, I thought.

But the twenty-eighth came and went, and Michelle showed no signs of going into labor. We meanwhile missed one of the perhaps two large parties a year that occur in town. No sign on the twenty-ninth. Finally, on the morning of the thirtieth, something began to happen. Michelle, who up to now had comported herself like any other cow, began to walk quite slowly and stiffly. My neighbor Floyd, who has known cows about forty years longer than I have, attributed this to overindulgence in pasture apples – no connection at all with the pregnancy, he said. But I had seen my wife walk the same way; I was sure it presaged labor.

Still, no dilation when I did my morning tail-check. None after lunch. That evening, though, just before my wife and I were supposed to be going out for the first time in a week, I noticed Michelle moving stiffly and heavily away from the herd. It is well known that cows like to be alone when they deliver. I dashed over, clean shirt and all, and found her partially dilated.

It was a dinner party we were going to, in another town, seven miles away. I found I didn't have the nerve to call my hostess and explain that I preferred to stay home and watch a cow. If Michelle had already started labor, I might have – but it could be another six hours before she did that. I resigned myself to missing the show, and to meeting a very young calf next morning.

Next morning I couldn't find either Michelle or the new calf. Ten of the thirteen other cattle were grazing quietly in the best part of the pasture – the rich, level section in front. Three were hanging around a section of fence further back, where my land adjoins Ellis Paige's, and doing some mutual head-and-neck licking over the fence with his Angus steers. But no trace of Michelle, even though I walked to the very back corner, where the pasture turns into pine trees on low ridges, with grassy glades in between.

About 8:00 A.M. I was plodding back toward the road, rather worried, and met Floyd walking briskly out. He likes births, too. Having just finished his morning milking, he was coming over to see how Michelle handled her first delivery. (He *knew* how I'd handle any medical emergencies: ineptly. That may even be one reason he came over.)

Floyd cheered me right up. In a rolling pasture like mine, he said, and especially one where there are lots of trees and bushes and glacial boulders the size of delivery trucks, you can some-times spend half a day finding a cow that wants to hide.

We began a systematic search, walking fifty feet apart, like a couple of corvettes hunting a submarine in tandem. About eight-thirty, the senior corvette made contact.

"There's your calf," said Floyd, pointing to a white spot on one of the pine ridges. "She's got it hid in them pines."

But as we walked quietly closer, it turned out to be a single white birch hidden among the pines; the sun was lighting up a patch of bark just about the size of a Hereford's head. We resumed patrol.

By nine o'clock we had searched the whole pasture except Bill Hill. Bill Hill is on the eastern side; it's about two hundred feet high, and practically vertical. There's maybe an acre of good pas-ture up on top.

I hadn't even considered looking up there, because I took it for granted that a 287-day pregnant cow would not be able to climb up, especially when she's got a dilated uterus and stiff legs. But there's nowhere else *to* look. We now clamber up, Floyd in the lead.

The pasture fence runs parallel to the top of Bill Hill, slightly down on the far side, and there are several bushy little glades right along the fence line. In one of these Michelle is lying. It's about as private a delivery room as you could ask for, and delivery has just begun. Michelle has already passed the water sack that helped to dilate her, and she is having contractions about every three min-utes. Each time she does, we can see two tiny light-colored spots appear. They are baby hooves.

All this we have been watching from a safe distance – thirty yards, I'd guess. Floyd now wants to get closer, to make sure those are the front hooves we're seeing. They'd better be. If they're the rear, we'll have to try to reach in and turn the calf. Otherwise there's a high risk it will be stillborn, since the umbilical cord will break before it emerges and can start breathing. Very quietly we move toward Michelle's glade. She watches us.

When we get halfway, she heaves herself to her feet. Still walk-ing in that slow, arthritic way (it *is* labor, not too many apples), she lumbers about a hundred feet down the fence line, and stands there looking at us. After five minutes, when we have come no closer, she lies down again in another little glade.

This time we circle around the top of the hill and come over the crest directly above her. We are within thirty feet, but shielded by a clump of fire thorns. She seems not to notice.

Another contraction, and the little dots appear. At this distance even I can recognize them as hooves, but it is still impossible to tell which pair they are. I am bitterly regretting that I didn't have the sense to shut the barway between the front and the back pastures last night before we went to the party. If I had, this scene would be taking place somewhere in the level front part. Suppose those are the rear hooves, and we wind up having to call Dr. Webster. He isn't going to be thrilled at the prospect of treating a patient on top of Bill Hill.

The contractions are beginning to show results. The little hooves now come out a couple of inches each time, and they remain visible between contractions. That happens five times. Michelle must be exhausted, because now she simply stops the whole process for about ten minutes, and rests. Chalk up one for instinctive animal powers.

Then there is a powerful contraction, and we see something white. A calf's nose! Those *are* the front hooves, and all is well. Another minute, and the whole head is out. Michelle rests briefly again, and then with a second powerful contraction pushes the calf halfway out.

At this minute Floyd gets up from where we are sitting quietly behind the fire thorn and starts walking directly toward her. "Floyd! You'll make her get up again," I hiss.

"Want to," he says, and keeps going. Slowly, but straight at her.

When he's about ten feet away, Michelle finishes bearing the calf with one mighty contraction, gets up, and backs away. We can now see that it's a bull calf (the females in the house will later name him Armand) with dark red curly hair. Already at birth he has a bull neck and the big clumsy ankles of a Hereford. He's charming and babyish – though a big baby, eighty pounds at least – but the only thing about him you'd call *delicate* is his little white head.

Floyd, still walking slowly, comes right up to Armand, who is now about thirty seconds old. He is lying just as he was born, his whole effort concentrated on trying to breathe. Not too successfully: he is rattling and wheezing. In my own childhood, human babies got spanked under these circumstances. What Floyd does is pluck a handful of dry grass, wipe the mucus from Armand's nostrils, and then knead his chest. The calf begins to breathe normally. He is one minute old.

We are not the only visitors who have been watching this birth. As Floyd turns to get another handful of grass, a series of dark spots appears on the little white head. Flies. Five of them. You'd think they might have given him one free hour, but that's not the sort of thing that occurs to flies.

Floyd has his new handful of grass. He quickly dries the calf's body, and then steps back to where I am. We watch anxiously to see if Michelle, that young and totally inexperienced mother, will own her child. Meanwhile, Armand takes his first independent action. He shakes his head briskly, and the flies depart. The motion attracts his mother, who comes over and begins to lick his rear end. God bless instinct. It's true she started at the wrong end – she should have begun with the head, and thus cleaned his nostrils – but this is still owning. At about the second lick, Armand moos for the first time – a soft baby moo – and Michelle answers. A few more licks and he has a bowel movement. He is four minutes old.

Michelle continues to ply her tongue, and as I watch I begin to understand that old phrase about licking something into shape. She is working gradually up toward his chest, and clearly it makes him feel good. He gives a little grunt, pushes with his front legs, and tries to stand up. He makes it halfway. But he hasn't learned how to coordinate all four legs at once yet, and he instantly falls down again. Michelle keeps licking.

By now it is ten-thirty, and Floyd and I both have a day's work ahead. And I, like the flies, am ready for breakfast. But we would like to see the calf on his legs. We decide to wait a few more minutes. Armand tries twice more to get up and twice more falls down. On the fourth attempt, he makes it. The general rule is that a healthy calf ought to be up within an hour. Armand has done it in twenty-six minutes. It takes him only two minutes more to find his mother's udder. All is well; Floyd and I start down the hill.

That afternoon, the entire human population of the farm – four people – climbs up the hill. The little glade where the birth took place is empty. There's not even a trace of the afterbirth, which Michelle has obviously eaten, as cows are supposed to. We go further back along the hill, my daughter Elisabeth in the lead. Suddenly she gives a cry of triumph. There in a thicket of young poplars is the calf, standing with his legs splayed out, perfectly motionless. Michelle is nowhere in sight. Presumably she's off grazing, and has hidden her son here, leaving him orders to stay perfectly still. *How* a cow conveys such an order, no one knows, but there is no doubt that she can.

Elisabeth tiptoes up and pats the small white head. The calf rolls his eyes, but otherwise stays perfectly motionless. "You darling thing," Elisabeth says. "I think I'm going to call you Armand."

Our farm is now a farm.

[1983]

76

Two Letters to Los Angeles

🎭

I

NOVEMBER 10, 1980 — Today it snowed six times in Vermont.
These are the fourth through ninth times it has snowed so far
this fall.

Before you start imagining pretty white flakes drifting down, let
me describe a few of today's snowfalls. The first two were mere flur-
ries, grayish, sleety snow that didn't even stick. The third was bigger, but no more romantic. It began about 11:00 A.M. I was out in
my woodlot, stacking red maple logs that I had cut and split yesterday. Promptness matters here. If I don't stack them within a day
or two, they freeze to the surface and are locked in until spring.

Quite abruptly, a sharp little wind came up, and then the ground
began to hiss. Snow pellets were falling thickly – not straight down
from the heavens, but blown at a twenty-degree angle from the east.
They didn't come hard enough to punch holes in the dead leaves,
or even to sting the face much. Just hard enough to hiss against the
ground.

When the squall stopped around eleven-thirty, the woods were
actually rather pretty. By early November the fallen leaves have
mostly lost their color – but in the right light you can still see
traces of their once-glorious reds and yellows and oranges. The
end of the squall produced a few thin gleams of sunlight that were
the right light. Most of the woods' floor was covered with a quarter-
inch layer of snow pellets, precisely that oyster color that some
women seem to prefer for painted woodwork. But in the lee of

77

each evergreen tree, and even behind the trunks of the larger maples, there was a patch of bare leaves, faintly glowing. Think of those as comparable to the windows in the oyster-colored houses. Though I was a little cold, and my leather gloves were sopping, I took an appreciative walk before I got back to stacking cordwood.

The afternoon was worse weather than the morning, at least for my purposes. Now out of the woods and back home, I was waiting for a neighbor to come and slaughter our lambs. Meanwhile, the sky had changed from alternating sun and snow to a steady, mean gray.

George arrived with his gun and his set of knives around 2:00 P.M. From then until dark, he was busy killing and skinning lambs, and I was busy salting down the hides, burying guts in the garden (terrific fertilizer), and helping him on two-person parts of the job. All of this would have been a good deal more pleasant if the temperature had not remained a constant thirty-three degrees, and if, at intervals, we hadn't got another handful of snow pellets flung in our faces. You can't skin lambs with gloves on; you can't even properly salt down hides that way.

Just before dark – that is, about twenty to five – we put the carcasses in the back of my pickup and took them over to Enoch Hill's. Enoch lives half a mile away, and is a country butcher. About a week from now he'll return me four legs of lamb, many packages of tiny lamb chops, and more sheep kidneys, hearts, and lungs than I ever know what to do with. I was so cold by then that I went home, crammed the main stove as full of logs as it will go, and went upstairs to take a hot bath.

This, of course, is autumn. We still have winter to come. Today's bit of snow will almost certainly melt, even on the north sides of buildings. But some time two weeks or a month from now we'll get snow that means to stay. We'll have it until the end of March. When it goes, we'll have more in April – big wet snows that stick. Once in a shady cove on the river, I found snow on the fourth of May.

Why on earth does anyone choose to live here? There are known to be options, including places where November might yield a string of languorous days in the eighties.

78

Enoch was born here, and in that sense maybe didn't choose. (Though he could always emigrate, as a God's plenty of his fore-bears did. There's a steep ridge not ten miles from our village which for more than a hundred years has been called California Hill. From its top, Vermonters setting out west in their covered wagons took a last look back at the frostbitten landscape they were leaving.)

But George and I were not born here. We both came fluttering into the state like moths, drawn to the light. So what is it that draws us?

My mother used to say it was masochism. She could understand coming for the brief, lush-green summers, but a will to spend the other seasons here she could only attribute to perversity.

Myself, I think there are two quite different reasons. One depends on the well-known theory of challenge and response. If you want to have responses, the theory says, you need challenges. These our rotten climate provides. I didn't really want it to be thirty-three degrees and snowing when we dressed off the lambs. In fact, I wondered out loud to George when he arrived if it wouldn't make sense to wait for a better day. He answered like a good Vermonter that (a) there might not *be* any better day, and (b) in any case best get it done now, while the knife was sharp. And once we had done it, and even though I only assisted, I felt a little bit heroic, as if there'd been an accomplishment. To feel that is an agreeable response.

The other reason is simply the human love of variety and per-haps even of unpredictability. This love is, of course, balanced by an almost-as-strong love of sameness and predictability, which is why uniform "national" products are so reassuring, why people stay at Holiday Inns, where, when you've seen one, you've seen them all, and so forth. But love of the unpredictable is a shade stronger, strong enough, sometimes, to keep people out of paradise.

I can best illustrate that with the story of a hitchhiker I once picked up. This was during mud season, an April afternoon when it was alternately snowing and raining. I was taking the back road between two remote villages, and came on a teenage boy with his thumb out. Of course I picked him up, and of course we got into talk. He seemed strangely cheerful for a kid in a wet blue-denim jacket out hitching in such miserable weather. When I eventually asked him where he was from, he said St. Thomas.

"St. Thomas? I thought that was one of the Virgin Islands."

"It is."

"But don't the Virgin Islands have a perfect climate, just like Los Angeles, never too hot and never too cold, and always sunny?"

"Yes."

"Can I ask you how you happened to come to Vermont for mud season?"

"Oh," he said casually, "I just got tired of one god-damned perfect day after another."

II

NOVEMBER 18, 1980 – The tenth snow of autumn was expected to be a mere squall, like the other nine. One pessimistic forecaster said there might be an inch or two.

But when we got up this morning, there was six inches on the ground. More snow was falling fast. That's not supposed to happen before Thanksgiving – and in fact it hasn't happened in about a decade.

All the stuff that careless people like me had left outdoors was buried. There weren't even mounds to show where axes and garden hoses lay – just this smooth white sheet. It was a true winter scene. All the evergreens drooped their branches under heavy loads of snow; all the roads were salted and hideous-slushy. School was canceled in the whole region.

My neighbor Floyd and I had planned to spend the morning loading the last of his manure pile in a spreader and spreading it in the pasture that our cows share. After that we were going to store the spreader in my barn for the winter. In the afternoon Floyd had meant to go get his November sawdust at the mill. (He uses it for bedding the cows.) Me, I had meant to buck up a big ash log at the edge of our shared pasture, split it, load it, and deliver it to a firewood customer in the next village. It was one I had cut in the spring, and leaf-dried.

Naturally all these plans got changed. Floyd makes part of his winter-living plowing driveways. Before dawn he spent an hour mounting his plow, and was on the road in time to plow out people who leave for work at seven. Then he did the morning milking and went out to plow again. I spent half an hour feeling around in the snow with my boot toes for tools. (I found them all.) I brought in a lot of snow-covered wood for the stove, and fed the beef cattle. Finally, around nine o'clock, I trudged into the woods to rescue my tractor. I had it parked about half a mile off the road.

That's not quite as foolish a place to keep it as it sounds. In fact, it's a very sensible place. Every year when we're finished cutting hay – this ought to be early July, but it is more usually around the first of September – I take the mowing machine off. Then I mount a hydraulic wood-splitter on the back and wiggle the tractor into whatever part of the woods I'm thinning that year. It stays there until just before I think we'll have a serious snow.

I was lucky. Even though the snow was now seven inches deep, and getting more serious by the minute, the tractor wiggled back out of the woods without getting stuck even once. Because I didn't dare take either hand off the wheel, I did get a few facefuls of snow, passing under hemlocks with especially low-drooped branches, but that was a cheap price for getting the tractor safely out.

What I should have done was take it right home, drive it into the barn, and mount my own snowplow blade. Then plow out my barnyard. But the ease with which I had wiggled through half a mile of woods made me overconfident – besides, there's a certain thrill to taking risks. What I actually did was drive straight to the pasture and over to that big ash log. I lowered the splitter near the butt end of it. Then I walked back out on the tracks, got my truck and chain-saw, and drove them in, too. Two hours later I had half a cord of beautiful ash firewood neatly stacked in the back of the truck. Meanwhile, the snow had begun to taper off (there was just under

81

nine inches on the ground), and the temperature had risen to a couple of degrees above freezing.

I thought I'd take the truck out first. It's four-wheel-drive and quite nimble. Even though the pasture is a very old one, and you enter it through a little curving lane lined with granite posts that some farmer quarried on the place around 1800, I didn't see why I couldn't boom on out. The heavy load would make for beautiful traction.

And so it would have, except for the unfortunate matter of its having got so warm. Snow at thirty-four degrees Fahrenheit, lying on frozen grass, makes something close to a friction-free surface. I did get halfway up the lane. I might even have made it if the truck hadn't begun to side-slip right into a granite post eight inches by eight inches and five feet tall. There was nothing to do but stop and get out.

A few years ago I would have panicked. I would have unloaded all the wood, right there in the snow. I would have walked home and gotten the kind of hand winch that we call a come-along. Then I would have spent the rest of the afternoon hand-winching. No one wants to have his truck and tractor trapped out in the pasture from mid-November until sometime in April.

But there are a few advantages to our climate, and one is that you can always count on it to get colder again. What I actually did was go do other things, and wait for dark. By 6:00 P.M. the temperature was well below freezing and the snow nicely crisped up. Truck and tractor came briskly on out – and their headlights on the untrodden pasture snow even had a kind of beauty.

At six-thirty I came crunching into the yard of my customer. He's a newcomer from the city, a former New York accountant who's been accounting in Vermont for just a couple of years. For that reason I called ahead to see if he really wanted his last load of wood to arrive on a snowy evening. Customers are expected to help unload.

"Sure, why not?" he'd said casually. "Wait any longer, and the cellar doors may be buried."

Two Letters to Los Angeles

He was waiting by the open bulkhead – off which, of course, he had just shoveled nine inches of snow. We began to pitch the wood down. I was wearing thermal boots, and overalls over my pants, and a wool jacket, and leather gauntlets. Snow-covered wood does get to feeling cold. But though he, too, had boots and a jacket, he hadn't bothered with gloves. With his white accountant's hands he gripped those icy logs and pitched them down cellar as lightly as if he'd been tossing grapefruits.

"I've got another pair of gloves in the truck," I said after a minute. "You want to borrow them?"

"Oh, hell no," he said. "It's only half a cord. And the weather's not very cold, anyway."

If this is how a city accountant behaves after two years, what will he be like after ten? Probably be getting sore if we don't have frosts in July. As for southern Californians, if any of them ever move here I firmly expect to see them riding down the hills on their surfboards, snow spraying up on both sides, dressed just as they would be at Hermosa Beach. Floyd and I will watch respectfully. But we won't join them.

[1980]

How to Farm Badly
(and Why You Should)
🍃

In one of Albert Payson Terhune's dog books, there is a background figure of a rich city man who has bought a farm. His first act is to stock it with prize cattle and pedigreed sheep. Then he buys a lot of expensive farm machinery. Finally, eyes shining, he sets out to improve the pastures (he wants to grow prize hay), build the best fences in the county, and in general turn his farm into a showplace.

Terhune wrote that account sixty years ago. But the tendency he describes is still very much in evidence. When city people buy an old farm, not just as a venue for lawn parties, but because they are converts to country life, they usually get carried away. They start to fix up the whole farm the way you might fix up the interior of an old house.

This impulse is highly understandable. I have a bad case of Improver's Itch myself. A month seldom goes by that I don't fix up something or other on my own farm, even if it's only rebuilding a couple of rods of stone wall. Over the years I've handled a surprising amount of stone that serves no farm purpose whatsoever, and hasn't since wire fencing came in. But I love doing it.

Furthermore, in the case of people who have bought a rundown old place, the impulse to fix is not only understandable but necessary. There are apt to be dead cars behind the barn. It makes every kind of sense to get them quickly off to a junk dealer. The fields are likely to be growing up to brush. The faster you start clearing, the better. Old fruit trees are sure to need pruning. The day you move in is not too soon to start – provided you know how to prune. All that I concede.

But the minute anyone starts thinking "showplace," he or she is inviting about six kinds of trouble. Unless like Terhune's Wall Street Farmer, you *want* to sink an annual fortune into the place – and unless like him you have a crew of hired hands to solve all the problems you are certain to create – you will be well advised to start out farming badly.

That statement needs immediate clarification. There is one kind of bad farming that is pure laziness or sheer ignorance, and that is a matter of not taking the one stitch now that will save nine later. In no way am I recommending it. Then there is a second kind that is bad farming only by the standards of agribusiness. What it requires is accepting your own limits and the limits of your land, and having the sense not to try for results that exceed those limits. This is what I am talking about.

Let me be as clear as possible. I don't mean letting a hillside field erode into gullies because you plowed it wrong. Certainly I don't mean getting a little flock of sheep and then watching them die of intestinal parasites because you didn't know about boluses and balling guns. I don't mean *anything* that harms the land.

What I do mean is avoiding most high technology. Not trying to achieve the "best" anything. Ignoring most (not all) advice put out by the U. S. Department of Agriculture. Or to put it positively, being content with moderate yields, modest improvements, slow changes, old equipment. Being content with this for the first five years, anyway, and probably even after that, unless you've meanwhile turned into a professional farmer. (In which case you may not be reading books like this, anyway, which you'll regard as sentimental. You'll be reading *Hog Farm Management* or *Agrichemical Age*.)

Enough of exhortation, though. Examples are what convince. Let me give some. They'll be Horrible Examples, of people who did try for the best, that person most usually being myself. Let me start with sheep.

Sheep have made a comeback in America since 1970. It is increasingly common for city people with country places to buy a few as their first venture into farming. Most do a little preliminary research – and quickly discover that they have a complicated choice to make. That's true with anything one looks at closely. When I was a boy in the suburbs, I thought there were three kinds of potatoes – or would have if I had ever bothered to concentrate on so boring a subject. I thought there were real potatoes, sweet potatoes, and something called an Idaho potato, which you baked. Since I've been a farmer, I've personally grown fifteen varieties of potato, including the purplish ones shaped like and called Cow Horns, and am aware there are hundreds of others.

So in the ovine world. The newcomer finds that there is not just one woolly animal called "sheep," as in the cute kids' ads, but instead breed after breed. Nearly all sound appealing: tall lordly Suffolks with their dark faces and rapid weight gains, Romneys that yield such tasty legs of lamb, Finn crosses that are so incredibly prolific. The newcomer sits around ticking off the advantages of each and wondering which to get, as one might compare gas mileage, comfort, and acceleration when buying a car.

The one thing the newcomer doesn't consider, usually, is getting an unpedigreed sheep. True, they're cheaper. (Last year a friend of mine got a sort of mongrel ewe at an auction for $22 – and within a week she had twin lambs, which makes three sheep at $7.33 each. But that was an exceptional case.) The difference isn't likely to be huge. You can usually get a good pedigreed lamb for sixty or seventy dollars, where you might pay thirty or forty dollars for a common animal. To a middle-class American, used to shelling out two or three thousand extra for a car that pleases him, what's thirty bucks? Go for the good stuff, he thinks.

Certainly it's what I thought when I got my first two lambs, which were Dorsets. What I failed to reflect on was that nearly all pedigreed sheep are highly specialized creatures. They've been

86

bred for maximum wool yield, or maximum meat yield, or maximum breeding speed. They've had that done to them for hundreds of generations. What human beings have cheerfully sacrificed on their behalf is versatility and resilience. All sheep die easily, but on the whole pedigreed sheep have an even feebler grasp on life than ordinary sheep. They're also more likely to need assistance in birthing. Owners of high-class sheep are often in the barn at 3:00 A.M. on a cold March night, helping to deliver high-class twins.

Neither of these, though, is the reason I wish now I had started with a couple of common lambs. My reason is grazing habits.

I had several aims in mind when I bought lambs. One, of course, was fresh lamb chops, organically raised. Another was sheepskins to put on car seats – it really does keep them cool – and daughters' beds. But my main motive was to acquire a mobile weed-trimming unit. I had a small orchard that I wanted to clear of weeds and brush without getting into plowing and reseeding. I imagined my two little Dorset ram lambs tirelessly chewing away at the hardhack and goldenrod.

They did, I admit, eat the poison ivy patch that had come into one corner of the orchard. And sometimes they would nibble daintily at young dandelions. Otherwise, they ate only grass, and only certain kinds of that, and only at certain early growth stages. I'm not blaming them. They had been bred to put on weight fast; and to do it at the rate they were genetically conditioned for, they needed the best high-protein grass. Plus grain on the side.

Low-bred sheep, on the other hand, are used to making do with what's available. They have their preferences, to be sure, but lacking first, second, and third choice, they can keep hide and hoofs together on almost any kind of pasture. They're what in New England used to be called thrifty keepers. They're what I should have bought. I might not have had any papers to show, but I would have gotten my orchard cleared. (Especially if I'd bought half a dozen lambs, instead of two.)

Time for another example. Following the Wall Street Farmer, let's turn now to land improvement. Almost anyone buying an old farm is likely to get a worn-out field or two with the place. The grass is thin; there are ferns coming in; whole patches are spotted with moss. Sure signs of acidity and poor soil.

There is a natural and healthy tendency to want to restore such fields to good condition: have some soil tests made, order the lime and fertilizer, maybe get some clover and vetch going. I'm for it. So is everybody else. The county agent may be able to show you how you can get the government to help pay for the lime. Any agricultural text, old or new, is full of instructions to fertilize. Even poetry points that way.

> *What is more accursed*
> *Than an impoverished soil, pale and metallic?*
> *What cries more to our kind for sympathy?*

says Robert Frost in the poem "Build Soil." Horace and Vergil held the same view.

So, fine. Lime and fertilize. Spread manure, if you have any. But if you take my advice, you'll do all these things in moderation. Try for good grass, but not exceptional grass – because if you have exceptional grass, it's going to require exceptional care. Again, an example from my own experience.

Behind my house there is a fairly good hayfield. Seven acres. It has only a few stones and no slopes too steep to mow. With a little lime every decade and a little manure every year, it consistently yields 250 to 300 bales of hay a year. I mow it, and a neighbor with more equipment comes and does the raking and baling for me. We need only two sunny days. If I start mowing on Tuesday morning as soon as the dew is off, he can rake on Wednesday afternoon and bale before supper. That night the hay is safely in the barn and it's nice bright dry hay. Then we let the grass recover for two or three weeks and turn calves in for the rest of the summer.

A decent seven-acre field *can*, of course, yield much more than 250 to 300 bales. Even in a single cutting, it can be made to yield twice that. Once a few years ago, when I had some horses to winter as well as my usual two or three beef cattle, I decided it would be silly to buy the extra hay when I could just as easily have a prize field. So I loaded it with chemical fertilizer.

The results were apparent right away. Grass that had been ten inches high the year before went up to twenty; and in the swales where it had been high to begin with, it looked something like a bamboo grove.

You can see what's coming. That year I harvested the smallest quantity of good hay I have ever gotten. The mowing went all right, though slower than usual because the swathes were so heavy, and I kept bunching it up at the turns. But after that I had a steady series of disasters. I knew that heavy hay would be hard to dry, and I had been able to borrow a hay conditioner (an old one that worked something like a laundry wringer). In the intervals between getting the conditioner jammed, I was able to wring the juice out of most of the really tall grass. That helped – or at least I'm determined to think it did – but it didn't help enough.

When my friend came to rake, he took one pitying look, and went home and got his tedder. In case you're not familiar with tedders, they're spidery-looking devices that turn thick grass over so that the sun can get at the bottom side. Then he waited a day and came back and raked. Some swathes were fully dry, and some weren't. On that third afternoon, the sun still holding, we baled about ninety bales of good hay – and about thirty more with green locks, that would probably go musty. The rest we left. I spent until dark with a pitchfork, fluffing up the windrows. What I accomplished was to enable the nice rain we had the next morning to soak in even better.

Eventually we did get it all baled, and there was a lot. Nearly five hundred bales. Of course I figure it cost me about two-fifty a bale for rained-on hay in a year when one could easily buy bright hay for one-fifty a bale. I had been trying to farm too well.

You can say that's just me. A competent farmer (with lots of equipment) would have managed better. So he would have. But, then, few newcomers *are* competent farmers in the first few years. Besides, I have more examples that are not personal.

89

A friend of mine – pretty good farmer, too – has a sideline of making cider. The farm he inherited had about a hundred old apple trees, most of them not worth saving, he decided. So he pulled out about seventy of them and planted new classy stock from a nursery. Most of them died, being specialized, delicate, high-yield creatures, much like pedigreed sheep, and beyond his skill to care for. Meanwhile, the thirty surviving oldsters, helped by a good pruning, went right on yielding quite a lot of apples. They weren't big, and they weren't beautiful, but they made wonderful cider. Not so with the culls he was getting from a nearby big and beautiful orchard while waiting for his trees to grow. He got a lot of juice out of them, but not much flavor. If he could resurrect the seventy tough old trees, he'd do it.

Well, maybe he wasn't smart, either, so here's another example, one that doesn't involve anyone I know. This is just about a dairy farmer I've heard of, who upgraded his herd of milkers.

Most dairy farmers in America now keep Holsteins. There are still Jerseys and Guernseys and milking shorthorns around, but Holsteins are the norm, for the simple reason that they are enormous cows with enormous udders that give enormous amounts of milk. There is, however, a price to pay for all this enormousness, and I don't just mean their enormous appetites.

Anyone who has seen a Holstein milker (many people haven't, because once grown, Holsteins are often confined in a barn for life) – any such person knows that here is an animal that has been bred into distortion. A Holstein cow is basically a support system for an udder. So much so that the biggest and "best" Holsteins walk rather the way camp followers used to when they were smuggling whisky to the troops during the Civil War. Which is how? Well, those were the days when dresses reached the ground. Men, even military police, knew it was highly improper to lift anyone's skirt. So the camp follower would put a pair of suspenders on under her dress – and from them she would suspend a five-gallon can of whisky, which hung between her legs. Naturally this gave her a somewhat waddling walk. That's more or less how Holstein cows look, once they have made bag.

An udder that big is, of course, in a constant state of tension. Most Holsteins are perpetually on the verge of getting mastitis, alias inflammation of the breast, which is one reason there are antibiotics in most cattle feed. Some of them have to wear a sort of horrible parody of a bra called a Tamm udder support. It fastens with a lot of straps across the cow's back. You can get your cow one for about fifty-five dollars.

The farmer I read about already had Holsteins. He just wanted ones with extra-big udders, which would yield an extra thousand or two pounds of milk a year. So he did a bit of genetic engineering. That is, he bought semen from a bull whose get were guaranteed to be even more distorted than most Holsteins.

Result: his next crop of heifers looked as if they were carrying *six* gallons of whisky between their legs. Or maybe eight. Too much, anyway, for flesh to take the strain. The result was the bovine equivalent of a hernia. It's called a prolapsed udder, and what happens is that the udder droops until it drags on the ground. Not all his new heifers had their udders tear loose – if I remember correctly, it was no more than one in four or five. But it was enough so that he'd have been further ahead if he'd farmed a little worse. Further ahead economically, I mean. With heroic restraint, I'm not even raising any of the moral questions I see involved here.

Well, it's human to make mistakes, and maybe that was just a dumb dairy farmer I read about. So for my last example, I'm going to use a case that doesn't involve people at all. It involves forage.

Corn, hay, and oats are the classic forage crops in this country, but there are many, many others. Millet, for example, and mangelwurzels, the giant beets that figure as a comic effect in the novels of P. G. Wodehouse. In the last generation, sudan grass and several kinds of sorghum have become increasingly popular forages with American farmers. They're often served as a kind of salad (the technical name is green-chop) to animals that aren't allowed out

THIRD PERSON RURAL

in the fields – milking Holsteins, say, or beef cattle in that special western hell known as the feedlot.

Guess what. If you grow sudan grass or sorghum and are content with a reasonable yield, you can give it freely to cattle. But if you really put the fertilizer to it, you will poison the cows. What happens is that the prussic acid always present in sudan grass and sorghum shoots up to toxic levels the minute you try for maximum production.

I think sudan grass and sorghum are trying to tell us something.

[1982]

Nuclear Disobedience

❦

THIS ESSAY IS going to be a little bit embarrassing in its present company, like a Jehovah's Witness who has strayed into an Episcopal picnic. He *will* preach. The Episcopalians may like what the fellow says, but he's too earnest for them: he keeps waving his arms and making the same point over and over, long after the entire audience has grasped it, committed it to memory, and possibly tried repeating it backwards in rhymed couplets. If only he'd sit down and eat a deviled egg, and stop all that shouting, they could get on with the day's activities.

As it happens, I am an Episcopalian myself, and very fond of picnics. Normally I would sympathize with the other essays and be on their side against this one. Even as it is, I won't blame them much if they crowd back along the page and glower. But for once I am constrained to play the sweating Witness.

What I have to witness is the familiar fact that the United States possesses weapons which are too powerful for it to control, and which may at any time destroy us and the world, without anyone's ever quite having meant to. We all know about our danger, and just as soon as our government and the Russian government (and, of course, the Chinese, French, and British governments) reach an agreement to disarm, we will all breathe a huge sigh of relief and maybe give up smoking. So we weren't to be extinguished after all.

Meanwhile, progress toward such an agreement is imperceptible, and the danger increases. What does any man do to avert it?

93

Well, some write letters to newspapers, and some distribute leaflets. Some go to see their congressman, and urge that the United States should renounce its nuclear bombs now, whether Russia does or not. (The congressman, if he is typical, explains that this would be bad politics.) A few daring ones sail their boats into the test areas or picket missile bases, and they are ignored or quietly put in jail. Most of us wait with a mixture of hope and resignation for our government to do something, and pray that extinction doesn't come first. And while we wait, we help to increase the danger. As Air Force officers, we fly live bombs over the Arctic, and sometimes over the towns where our children lie sleeping. As physicists we design new and worse weapons. As technicians we build them. As administrators we plan them. As taxpayers we pay for them. And we don't know what else we can do. For surely if there were anything, our government would tell us, or the people would rise with a thunderous voice and tell the government.

The worst of it is that those of us who write the letters and plead with the congressmen actually have a feeling of virtue. We tell ourselves that we are doing all a single man can do, and if we die in a nuclear blast it won't be our fault. Some of us think in our heart of hearts that whatever happens to the others, we won't die in one – it would be too unfair. At the last minute, we secretly feel, some god will step out of the machine and rescue those of us who protested. Or at least one ought to.

Henry Thoreau, from whose essay "Civil Disobedience" I take my text, has something to say about this feeling. He was talking, a hundred and twelve years ago, about those Americans who knew in their souls that slavery was wrong and who wished to see it ended. "They hesitate and they regret," says Thoreau, "and sometimes they petition; but they do nothing in earnest and with effect. They will wait, well disposed, for others to remedy the evil, that they may no longer have it to regret. At most, they give only a cheap vote, and a feeble countenance and God-speed, to the right, as it goes by them." Such as these, says Thoreau, "command no more respect than men of straw or a lump of dirt." So much for our sense of virtue in that we wrote a letter or signed a petition. Men of flesh have to take stronger action than that.

There's another problem, of course, and Thoreau deals with that, too. It is hard for a single person *to* take much action, in a country like the United States. Solitary action seems undemocratic. As Thoreau puts it, "Men, generally, under such a government as this, think that they ought to wait until they have persuaded the majority...." If a minority of us know that we must renounce nuclear weapons here and now, while we still can, and the majority hasn't realized it yet, then our job is to educate and persuade the majority. And how are we to do it, except with letters and petitions and television shows, and other harmless expressions of opinion?

On all expedient matters, Thoreau would agree with this view, and so must any good citizen. On matters of total conscience, such as slavery and the use of radiation, another and a harder rule applies. In matters of total conscience, men sometimes have to disobey the government and the half-felt will of the majority. Indeed, the disobedience of conscientious men may provide the only means through which the majority can find its true will. The thunderous voice of the people has its origin in the stubborn throats of just such men. Silence them, and there is no check left on government but the opinion poll, which is no check at all. Thoreau puts the case more succinctly. In matters of total conscience, he says, "Any man more right than his neighbors constitutes a majority of one already." As this special kind of majority, it is his plain duty to act.

What this means in the United States now, it seems to me, is that those who care whether humanity survives must begin to risk something more than their signatures on a petition. Those of us who fly live bombs could always try refusing. Those of us who build them could look for other work. Those of us who are reservists in the armed forces – and I am one myself – could serve notice that we will not fight in a nuclear war. (That very few of us would have a chance to fight – our chief role, like that of other people, being to perish – is for the moment beside the question. So is the fact that we seem just as likely to end our species in a peaceful accident as in war.) Those of us who finance this petard with which we are to hoist ourselves could even try not paying our taxes. It would be interesting to see what happened if two or three hundred thousand of us did refuse to pay next year.

There would be a special rightness in Americans doing these things. As much as anyone, we are responsible for letting the weapon out of control in the first place. Our technology and our genius built it; our money paid for it. We were the ones who took an atomic bomb which, in order to serve warning on Japan, we could perhaps have dropped in the open sea off Yokohama or into one of the great inland forests, and released its radiation onto a city full of human beings. Three days later, while the Japanese were deciding whether or not to surrender, we repeated the act on another city. Japanese are still dying of leukemia as a result. There is a chance that for the rest of history some Japanese babies will be monstrously mutated as a result. We did that.

That the Japanese would almost certainly have loosed radiation on our cities in 1945, if they had had the bombs, is no counter-argument. We are the ones who did do it, and in consequence we have a little more atomic responsibility than anyone else. Decent Germans must feel a special concern for Israel because of the Jews Germany slaughtered, and decent Israelis must be concerned for Palestine Arabs because of the land Israel has taken. Decent Americans must feel that concern for the whole human race, insofar as we have threatened its health and its survival with our free use of radiation. Possibly we had to do what we did in 1945. Possibly we have to be the ones to stop now. Even self-interest suggests that. After all, anyone who believes the Japanese would have used nuclear weapons on us in 1945, supposing they'd had them, must believe that some day the English, the French, the Russians, the Cubans, our surviving Indians *will* use them on us. Anyone who believes that and who does not push in earnest and with effect toward disarmament is a fool.

One more point needs to be brought up, and I want to beat the reader to it. The point is simply whether all this talk about extinction unless the world gives up nuclear weapons isn't rather alarmist. After all, we Episcopalians have been going on our picnics for years, and we haven't been washed out yet. People have always been claiming that the world was about to come to an end, unless this or that was done, and they have been wrong every single time. Our government assures us they are wrong this time – and would probably add that those who refuse to fly live bombs or pay their taxes will most assuredly go to jail.

I like to imagine a council of Blackfoot Indians about the year 1800. They are discussing a rumor that white men are slowly moving west, and that they have with them a terrible new weapon that shoots fire. Certain alarmists on the council predict disaster.

"Pooh," answer the rest. "People said that when the bow-and-arrow was invented. Remember when those other white men came up from Mexico on horses? We had never seen horses, and you hysterical types were running around moaning that all was lost. Remember, we told you we'd get our own horses and restore the balance of power? Well, didn't we? Don't be so excitable. You'll be predicting the end of buffalo next."

Ask the surviving Blackfoot Indians whether or not their world came to an end.

I also like to imagine an informal conclave of Neanderthal hunters about the year 74,000 B.C. They are discussing a new kind of flint spear used by the Cro-Magnons in the most recent fight. Certain alarmists among the hunters predict disaster.

"Pooh," answer the rest. "People said that when the throwing stick was invented. We'll get our own flint spears and restore the balance of power. Too dangerous? You'd rather make a treaty? Listen, we'd rather run a little risk than make a treaty with those damn Cro-Magnons. What do you want to do, compromise the Neanderthal way of life? You'll be predicting the end of mammoths next."

As the Neanderthals were entirely wiped out ("Evidence from

Krapina in Croatia," wrote Professor Hooton of Harvard, "indicates in no uncertain terms that the Neanderthaloids in this region were eaten by their more highly evolved successors") ... since Neanderthals are extinct, it is difficult to question them. But ask their ghosts whether or not their world came to an end.

The only difference now is that with radiation we can all die together, instead of some doing the wiping out and some the surviving. Or even if there should be survivors of the nuclear war or the nuclear mistake, what guarantee has anyone that America will be cast in the role of the Cro-Magnons?

It is very easy to assume that government – ours, the Russian, the World Court, any government – must be right. Government represents legitimacy, tradition, law and order, the sanction of things as they are. These are things to be respected. And yet hear Thoreau once more. "A common and natural result of an undue respect for law is, that you may see a file of soldiers, colonel, captain, corporal, privates, powder-monkeys and all, marching in admirable order over hill and dale to the wars, against their wills, ay, against their common sense and consciences, which makes it very steep marching indeed."

It would be nice to hear that those against whom we march were abandoning nuclear weapons of their own accord, without waiting for us. But suppose they don't? Suppose they need the example of the United States, which our government, busy marching over hill and dale, seems unable to give them.

If a few of us who know the peril do not step out of that file, even if it means losing our corporal's stripes, who will there be to head off the column from the cliff?

[1961]

Postscript, 1983. The little essay you've just read is a genuine antique. I wrote it twenty-two years ago. It originally appeared as the final piece of my long-ago, long-forgotten first book. I haven't changed a word.

The terrifying thing is how timely it remains. Only two details are dated. No one is sailing into nuclear test areas these days, since the tests are all underground now. That's a gain for protestors, since there is less radioactive fallout. It's also a gain for the government, since an underground test is harder *to* protest, and also less conspicuous than devastating Bikini or Eniwetok, and hence less likely to alarm people. And the list of nuclear powers now goes well beyond the United States, Russia, England, France, and China. India, Israel, probably Pakistan, probably South Africa have come crashing into the club. There may be still more with a bomb or two.

Otherwise, things remain very much as they were in 1961. Politicians, ours and theirs, are still saying the same things. Eternally aimed nuclear weapons still point to Moscow and Washington, and probably hundreds or even thousands of other places as well. The danger of nuclear war has slowly but steadily risen, along with the number of weapons and powers. All of that is depressing.

And yet there is good news, too. World opinion is much more mobilized than it was in 1961. Back then, there was really no chance that people who understood the dangers could turn the course of things. There were too few of them, and they were too disorganized. Now there is a chance.

I can illustrate the change from my own life. Back in 1961, the only thing I could think to do for the sake of peace (besides write the essay) was to go on a "peace walk," one of the first ever held at Dartmouth College

The walk was organized by an undergraduate named Anthony Graham-White and by a couple of faculty members. We were to walk five miles. From Webster Hall, on the edge of the Dartmouth green, down to White River Junction, Vermont. Despite a few banners and signs, we were not an impressive sight. In those days Dartmouth had about 3,000 students and 250 faculty members, not counting the medical school, engineering school, and business school. Even though it was a beautiful mild day, a bare fifty people gathered for the walk – and some of those were neither faculty nor student, but townspeople. One was my infant daughter, in a baby carriage.

We were not the only group in front of Webster Hall. Peace walks, at least in those days, brought out war supporters, too. About twenty members of the Dartmouth chapter of Young Americans for Freedom were circling around us, waving *their* signs. The one that made a lasting impression on me was a big piece of yellow cardboard, with a rather good sketch of a mushroom cloud on it, and the caption "Keep America Safe." I wondered if the student artist who drew it realized it would work as well for our side as his.

We watched the Young Americans, and they watched us; nobody else paid much attention at all. Two fraternities were having a baseball game on the green, and a couple of hundred students were watching that, with maybe one bored glance at the milling little groups in front of Webster Hall. At the time, that infuriated me. I thought, "That's why we'll *have* a nuclear war. Except for a little group of fanatics who started it, and another little group of fanatics who tried in vain to prevent it, everyone was watching baseball."

With twenty years of perspective, I see the matter differently, see that apathy is by no means always bad. If everyone were "involved" or "concerned" all the time, the insanity rate would be up around 80 percent. There are so many causes and needs and injustices in the world that to let oneself care about even all the urgent ones would lead most of us to instant emotional bankruptcy.

All the same, averting nuclear war is a special case, since practically all other causes will cease to exist if we do have one. There may be survivors, but there certainly is not going to be much concern, the year after the war, over which party controls the Senate, or whether school instruction in New York should be bilingual, or how to raise money for the Los Angeles Philharmonic, or what progress is being made in curing cerebral palsy.

For that reason I rejoice at how much more involved people at Dartmouth are now. Last year, so many students and faculty went to New York for the great anti-nuclear rally that Vermont Transit ran out of buses. (And couldn't borrow any from other New England bus lines, because they had all run out, too.) This year there is an official part of the college, complete with office and staff support, called the Program for Education on the Threat of Nuclear War. It is the only program we have that's devoted to a single issue, and also the only one that in its name takes a position. The other programs are all called things like Policy Studies and Women's Studies and Environmental Studies. No one even considered calling this one Nuclear Studies, still less the Program for Education on the Threats and Blessings of Nuclear War.

Similarly, the only stand I'm aware of that the trustees have ever taken on a public matter was their statement in April, 1982, urging both the college and the country to learn more about nuclear perils.

These are encouraging signs, and they are to be found everywhere. In 1961, little Vermont towns, far from any college, considered it their business at town meeting to mind the town's business. By 1981, some of them felt differently. Eighteen voted, along with setting a highway budget and electing someone to be selectman, to request the federal government to cease both the testing and the production of nuclear weapons. That's only a tiny fraction of the 246 towns in the state; it was also only a tiny hint of what was to come. In 1982, another 161 towns passed the same resolution. In West Windsor it passed by unanimous voice vote – and then the whole town meeting rose and sang "America."

But the best news is that in 1983 there is good solid organization for the kind of resistance it will take to move politicians and generals. Non-binding votes in little Vermont towns certainly aren't going to. Or even non-binding votes in big states.

Back in 1961, I wistfully speculated on tax resistance. "It would be interesting to see what happened if two or three hundred thousand of us did refuse to pay next year," I wrote – and of course went on to pay my 1962 income tax, telephone excise tax, gasoline tax, etc. Apart from all other considerations, it is a scary thing to start refusing to pay taxes all by oneself. There were no fellow refusers to give me courage – or if there were I didn't know about them.

Vietnam has happened since then, and there is now a large group of Americans trained in what might be called loyal disobedience. Maybe even patriotic disobedience. Some of that training was in refusal to pay taxes. It really isn't all that scary once you get into it. Since 1983 there has been an organization whose members have all signed a pledge to begin weapons-tax refusal as soon as the membership reaches 100,000. It's called Conscience & Military Tax Campaign, and I belong.

The threat of nuclear war has clearly increased since 1961. But the will to resist preparations for the war has clearly increased even more. I think it is in the process of becoming a national movement. Farms and farmers may survive yet, along with symphony orchestras and medical research. That would be nice.

Last Person Rural

A Truck with Pull

⚜

I T W A S N ' T A big firewood truck, but it wasn't tiny, either. Maybe you've seen one like it: a three-quarter or one-ton pickup that's been converted to a flatbed, so that it can deliver two cords at a time.

This one had both its right wheels in the ditch, a muddy ditch on the side of a back dirt road in a back Vermont town. Probably the driver had moved over six inches too far, trying to let someone past the other way.

"Need some help?" I called.

The driver, a heavy, bearded man in his thirties, eyed my little blue Toyota pickup. Perhaps he also noticed my white hair. "Yeah, thanks," he said. "Probably what I better do is go get my skidder."

"If you like. But I think I can pull you out."

Then I pointed to what he *hadn't* noticed: the electric winch mounted on the front of my little truck. He almost smiled.

"Maybe you could," he said. "Be nice. It'd save me 'bout eight miles."

So I drove on past him, turned around, and came back, staying as far on the other side of the road as I dared. I stopped about thirty feet short of him, carefully aiming the nose of my truck at the nose of his. Then I got out the four chocks I keep behind the seat. By now he was in the spirit of the thing – he took them and chocked all four of my wheels. Meanwhile, I rooted around for the winch control, which also lives behind the seat (along with a six-foot logging chain, a pair of work gloves, and, effete touch, a large black umbrella.)

Leaving my engine running and the hand break set, I began to winch. It wasn't even hard. The winch groaned some, but it pulled him and his wood right back onto the road. Total time, including conversation: about seven minutes. After his ritual offer to pay and my equally ritual refusal, we went our ways.

My present truck is the fifth one I've owned, the second that was four-wheel drive, and the first with a winch. I can't imagine why I waited so long. I haven't had so much fun since I was a small boy playing with a toy steamshovel.

My new plaything is expensive. That I freely admit. When I bought the Toyota in the summer of 1987, I paid $13,000. More than $2,000 of that was for the winch.

Has it paid for itself? Of course not. I don't charge for my road services, and there haven't been many of them, anyway. Just that firewood truck, and once my sister-in-law's Bronco during an ice storm, and my daughter's car on the same hill during that same storm, and a friend's tractor in a boggy hayfield three years ago, and my own tractor last summer (I could hardly charge myself), and maybe two more. No profit there.

Only once has the winch produced a cash benefit of any size. That was when I had to take down a big dead maple last year. There's a row of old maples along the road in front of my house; back when they were planted it was a good place for a maple to be. But now they suffer from their proximity to road salt. They've been slowly weakening for twenty years, despite fertilizer, love, and – for a decade now – immunity from tapping. Last year one of the two biggest finally died.

It was not an easy tree to take down. If it fell back into the yard, the top would smash into my house. If it fell across the road, it would take the power lines and the phone lines. I've already knocked a power line down with a tree once in my life, and it's not an experience I'm anxious to repeat.

On the other hand, if this wide-crowned old tree fell to either side, it was sure to lodge in one of its neighbor maples, and a fine mess *that* would be. The only way that maple could safely fall was at an angle of about forty degrees in from the road, which would put the top near but not actually on my barn.

But how to accomplish that? Dead trees are much harder to take down than live ones, because instead of gradually bending on a hinge of green wood, they tend to snap at some unguessable point when you've almost cut through. Then they fall any way they please.

If I hadn't had an electric winch – and one with a hundred feet of cable, at that – I would have paid a tree company to come cut it. They, mindful of the house and the wire, would probably have cut it in sections, from the top down. It would have cost me $300, $400, maybe even $500. Instead, I just got a ladder and put a chain around the tree about twenty-five feet up. I parked the truck a hundred feet away at the forty-degree angle, and hooked on the cable. Then a neighbor ran the winch while I sawed. The tree landed within a foot of where I intended it to. Garrett, the neighbor, had brought his own chainsaw; he stayed and helped me buck the tree up. Total cost of removal: $0.00. Fair credit for winch: $400.

That still leaves $1,600 to amortize – and as I've already admitted, I can't. Not in cash. Try me on pleasure, though, and I might get as high as the equivalent joy of two weeks in a good hotel in Paris. Here are some of the things that keep me busy and happy with my winch.

My farm – my sort-of-farm, I should say – has one old pasture with a lot of well-rooted brush in it. Big mean stuff, often with thorns. The barberry bushes are, of course, the worst. Even yanking them out with truck and chain did not always work. I'd get the chain wrapped, hop in the cab, put the truck in first gear, and slam forward. One time in three or four, the bush would pull out, roots intact. Then the truck and I would shake hands, so to speak. More usually, the chain would slide up over the bush, and come loose in a shower of leaves. Or else the main root would snap. With the oldest and toughest bushes, nothing would happen at all except a nasty shock to the truck frame when the chain drew tight.

The winch has changed all that. With a winch one has exquisite control. I can tighten the chain so slowly that it seldom gets a chance to slip. If I do see it start to, I'm right there, I'm not in the cab of the truck. I can reposition it in about two seconds. With the hundred-foot cable I can pull barberries and firethorns off slopes where I'd never dare take the truck itself. Slope-picking requires a crew of two, of course: one down at the truck to run the control and one up at the bush setting chain. My wife loves winching as much as I do (she once said that if it weren't such a sexist phrase, she'd be glad to be called the winch wench), and on slope days she comes along with me. I don't mean to sound as if there were one of these every week. There have been maybe four in all.

Another thing the winch is good at is snaking logs out of inaccessible places. My farm is a hilly one, and the wooded parts are the hilliest of all. Some of its woody slopes are below field level, rather than above. It's quite handy to drive to the top of a bank that I have never gone down except on foot, hook on to a nice straight pine log forty feet below, and soon have that log on the pile, ready for the traveling sawmill that comes by every year or two.

But my favorite winching sport is moving entire trees. Making Birnam Wood come to Dunsinane, you might say.

The same rundown pasture that has all the thorn bushes also has a lot of bull pines in it. (Bull pine: a pine tree that grew up in an open field, and that therefore has huge branches right down to the ground. Sometimes called a wolf tree.) Years ago I took most of the little easy ones out. The few big ones that had a good log in them have been sawed up and turned to lumber. What that leaves are the twisted, knotted, weevil-bent big ones. Once or twice a year I get ambitious and decide to spend a day playing matador to a few bull pines.

Pine cuts almost like butter, if you have a well-sharpened chainsaw. Felling one of those trees takes only a few minutes. And even though it may have as many as fifty big limbs, cutting it up isn't going to take much over an hour. Well, a little more in my case, because I like to get something tangible out of the job (besides another three hundred square feet of pasture where grass will now grow). So any straight branch between about two and four inches in diameter gets converted into pieces of sugar wood: fuel for the little evaporator we use to make maple syrup. Still, even with that extra cutting and loading, two hours at the most.

But here's the problem. What I'm left with in the pasture is something like a whale after the *Pequod* has got through with it. The trunk and 95 percent of the branches of this enormous pine tree lie flopped across my pasture. If I waited ten or fifteen years, the branches would melt in by themselves, and the trunk would surely go in a century.

But I'm the impatient type; I want that stuff out *now*. I have two choices. I can make a huge brushpile, let it dry for a year, and burn it. The trunk won't burn, just char, but I can move that later, section by section, with the tractor. Or, right now, I can load all those branches onto my pickup, and drive a couple of hundred feet over to the fenceline. Then I can pitch them all over the stone wall into the edge of the woods. I used to do that. The branches of a big bull pine make something like eight truckloads.

But not any more I don't. Now I just reel out my winch cable and drag the whole tree over to the fence. Then I cut it up there, stopping at intervals to pitch branches over. Macho experiences are not something I usually seek. Macho experiences remind me too much of war. But I have to admit that it gives me a delicious power feeling to see a very large pine tree reluctantly moving inch by inch away from its native stump and over to the place where I want it.

Are there then no faults at all with the winch, no limits at all to its power or my pleasure? Of course there are. All of the above. The most obvious limit is that the winch is rated at only 8,000 pounds – and even to get that you have to be pulling in perfect alignment with the direction your truck is pointed. That's why I was so careful to aim straight at the firewood truck.

The resistance of a large bull pine may well be more than the winch can cope with, especially if a couple of branches broke when the tree fell, and the stubs are jammed into the ground. I've often cut a bull pine in two before winching, and I always trim stubs.

Another problem: The little blue truck is my commuter vehicle as well as what I take into the woods, and the winch is definitely *not* a convenience when I park in town. It sits up higher than the bumpers of most cars, and protrudes beyond mine. Were I to bump into another vehicle while parking, there would almost certainly be a crunching noise as the hook of my winch made vigorous contact with the rear of the other vehicle. This makes me an amazingly cautious parker.

Winches are dangerous, too. If that cable ever snaps, I would not care to be in its path. That's one reason I'm glad the winch control is on a ten-foot cord, so that I can and do take it around to the far side of the truck when I pull anything heavy.

These are trifles. Mainly the winch is pure joy. It may even have made me a nicer person. I mean, how many people do you know who actively look for cars stuck on back roads, so they can offer free assistance?

[1991]

A Vermont Christmas

CHRISTMAS IS SUPPOSED to be snowy in Vermont, maybe even snowbound. It was not that way last year. The weather clearly intended to let us down. After every December snowfall – there were only four, and they were small ones – it promptly rained. On Christmas Eve, in the little town of Barnet, we had cloudy skies and maybe an inch of slippery mush on the ground. The roads were bare and muddy.

There were five of us in the farmhouse on Crow Hill: two parents, a grown daughter, an eleven-year-old son, and a grandmother. Out in the barn were only two horses and eleven hens. We had sold our sheep and planned to get new lambs in the spring. Like nearly everyone else in Barnet, we had a great deal of getting ready for Christmas still to do.

By ten o'clock on the morning of Christmas Eve we had done the regular chores and a good deal more besides. The horses were hayed and watered, the hens fed and the eggs gathered. There was a three-day supply of newly split dry stovewood on the front porch. Amy, the grown daughter, had taken Marek, the eleven-year-old, into the big town of St. Johnsbury (eight thousand people) to finish his Christmas shopping. Anne, the mother, had made and iced a giant cake. Now the true Christmas events began.

The first step was the wreath. The grandmother had had her eye for days on a big spruce tree at the edge of the woods above the house. Taking a pair of clippers (to cut small branches) and her son-in-law (to hold them for her, once cut) she sallied out. Her daughter meanwhile cut bunches of red berries from the elderberry bush in the upper pasture. Both tree and bush actually gained from the light pruning. And by eleven o'clock the grandmother had made a spruce wreath with red berries that would probably fetch twenty dollars in any city store. Mother and daughter decided jointly not to hang it on the front door. Who besides the snowplow driver would see it on the next-to-last house on the remotest road on Crow Hill? Instead they put it inside the front hall, where everybody stopping to put boots on or take them off would see it.

Next came the tree. The mother had had her eye on *that* for a whole year. It was a furry young spruce that had come up in a corner of the lower pasture. Sooner or later the father would take it down anyway, since what you want in a pasture is grass, not evergreens. Right now it was a graceful young tree just about the height of the mother herself – that is, five feet five inches.

For a Christmas tree, the whole family goes out. Marek carried the bucksaw and cut the tree; his parents carried it back between them. Three dogs formed an escort: the two family dogs and the dog from the last house on the road, who found life more interesting down here with the horses and the chickens and the boy.

We hardly had the tree up when two things happened simultaneously. First, the already dark sky grew darker, and it began to snow. And then the phone rang for Marek. The call was from two cousins known to him as Sam 'n' Eli. They live a thousand miles away with their mother, but spend their vacations in Barnet. Their father lives two miles up the road with his second wife, two stepdaughters, and a baby. That place is a sheep farm, too.

Marek, who had thought he would probably die unless he put all the really important decorations on the tree himself, now realized that a far likelier cause of death would be if he didn't see Sam 'n' Eli immediately. Amy drove him over through the now fast-whirling snow. And decided, since she was out anyway, to go once more into St. Johnsbury (it's only five miles), where there are stores and public Christmas decorations and even a small amount of city grime.

The rest of us went on decorating the tree, occasionally sneaking off to wrap another package. Our plan was to finish the tree, take a walk in the woods to enjoy the new snow, have an early supper, and then go to the seven P.M. carol service at the Congregational church in Lower Waterford. Midnight services sound wonderful, but in farm country seven P.M. draws a much better crowd.

The plan did not work out, because soon the snow thickened. The first truly heavy fall of the year came sliding down. One result was that Amy's car did not make it back from St. Johnsbury. Crow Hill is very steep, and its roads are full of curves. With three inches of wet snow, you almost inevitably slip into the ditch if you have a two-wheel drive car.

Amy had a good walk home in sneakers. Her father took the little farm truck with the winch and winched her out, but they made no effort to bring her car back up to the farm. They left it in a neighbor's level pasture, down by the brook called the Water Andric. Everything was now pure white, and the road the purest and whitest of all. No car had been up Crow Hill for hours, and even the tracks of my little truck soon faded. In places where there were fields beside the road, but no fence, it was beginning to be hard to tell where the road *was*. Not a night to take grandmothers on back roads to Lower Waterford.

It turned out to be a good thing we didn't go. Because about eight P.M. there was suddenly singing outside, and the light of candles. The five cousins from up the road and their parents were standing outside in the fast-falling snow. Everyone but the baby held a candle, and all but the baby were singing "O Little Town of Bethlehem." We hadn't heard them come, because there is a very sharp uphill turn where the end of our driveway meets the town road, and even though they had come in a four-wheel Bronco, they had gotten stuck. Never mind that now. We had them in for hot cider and more carols.

Then it was too beautiful to stay in. We found more candles. Eleven people and three dogs walked through the silent snow to the last house on the road – the grandmother stayed home – and sang "Good King Wenceslaus." The eleven-year-old, who had been allowed to open one present early, and who had headed unerringly for his new bugle (he feels packages, and we suspect him of having a hidden x-ray machine) blew a sort of fanfare. It didn't sound too bad in the soft snow. Then we got the farm truck and winched the cousins' Bronco back onto the glistening road. Tomorrow would be as white as Christmas can get.

[1989]

The Lesson of the Bolt Weevils

PROBABLY THE SHARPEST QUESTION facing an environmentalist in the 1990s is what to do about the law. Does a good environmentalist stay law-abiding and work within the system? Obviously the traditional answer is yes, of course. That's what democracy means.

But what if the system seems to be corrupt? What if the very people in the system who are charged with protecting the environment are the ones who betray it? What if the Forest Service itself builds roads at taxpayer expense in order to facilitate the clear-cutting of great swaths of our national forests? What if high officials of the Environmental Protection Agency simply announce that they've decided not to enforce the laws the environmentalists got passed? *This* deadline for atmospheric clean-up they've unilaterally decided to extend for three years; *that* regulation for strip-mining they've decided to ignore indefinitely.

Does the good environmentalist sigh a little and start gathering signatures for a new petition? Go back to court yet again? Or does he or she turn militant and stage a sit-in in the national forest? Maybe even drive spikes into the great, helpless trees, so that when the loggers come, their chainsaws will chatter to a halt?

It's a hard decision, and one the environmental movement has by no means come to a consensus about. At one extreme are those who say that ecosabotage is criminal behavior, to be treated like any other crime. And not only criminal, but counterproductive –

it forfeits public sympathy. Dangerous, too. Hidden spikes may endanger loggers' lives. At least once, they already have.

At the other extreme are those who say that ecosabotage is the only way left, at this eleventh hour, to save our planet. Yes, it's dangerous, but it's also dramatic. So far from forfeiting general sympathy, the ecosaboteur brings it into being. Spiked trees get the public's attention in a way that petitions never have and never will. Court cases have their place – a major one – but what we as a people are really moved by is stories, not lawyers' arguments.

They must be true stories, too; they must really have happened. There are hundreds of novels about nuclear disaster, some of them masterpieces, like *A Canticle for Leibowitz*. There are dozens of movies. None has had even a fraction of the conversionary force that Chernobyl did, or Three Mile Island.

Or think back to abolition days. Abolitionists petitioned and held public meetings for thirty years. What finally aroused the nation was a criminal act. John Brown, that abo-saboteur, had to raid the federal arsenal at Harper's Ferry, and be caught, and be put to death before his fellow Yankees moved against slavery in real earnest.

Not only that, some environmentalists insist, ecosabotage often *is* productive. The spikes will often save the actual trees they're driven into. The timber corporation will be disgusted at the costly delays. It will be fearful of all the publicity now attached to its turning a stretch of national forest into a sorry mass of stumps. It will withdraw its crews.

What continues to surprise me is that neither side in this long debate seems to look much at the one major example of ecosabotage we have had so far in the United States. A pity. Because it offers profound lessons for both sides. And, of course, an exciting story, too.

That story begins in Minnesota in 1972. The opening scene is quiet enough: we see a small and very private meeting. Officials of two Minnesota power companies (both technically co-ops) have gathered to plan a giant new coal-burning power plant. It's going to be a pretty good polluter. They decide to put it in the middle of North Dakota, next to the strip mine that will feed it.

They will then build a 430-mile power line, cutting across North Dakota and Minnesota to just outside the suburbs of Minneapolis. It will be the largest direct-current line in the United States, carrying 800,000 volts of electricity. It will run on a strip of land 160 feet wide. A strip 160 feet by 430 miles will consume about 9,000 acres of land. Along that strip the companies will erect a long row of steel towers – just under 1,700 of them. Each will be 180 feet high. At the very top the big wires will run.

Almost no one likes to live under high-voltage lines. So how do you persuade them to? The answer, if you're a utility company, is that you don't have to. You can force them to. Utilities are generally able to borrow the government's power of eminent domain. They send the appropriate agencies a projection showing that the power will be needed. After a number of hearings, approval is usually forthcoming.

That doesn't mean there won't be loud outcries from those who wind up under the line. In a democracy, such cries are expected. To counter their effect, the Minnesota companies decide to site the line "objectively." They hire an out-of-state firm to do it by computer. The computer is given a point system to work with – the more points a piece of land is awarded, the less danger of its having a power line across it. Land owned by the state of Minnesota gets five points, woodland is worth three points, and so on. Farmland, however, gets no points at all. Zero. Naturally, the computer plots the line, often diagonally, across one farm after another. Many of the farms will be cut in two.

The farmers, of course, know nothing about these plans – not even that there's going to *be* a power line. There's a good case for power-company secrecy, too. If word got out, wouldn't speculators be buying up possible right-of-way land, hoping to get a high price? You bet.

Two years go by. The power companies complete their plans.

They ask for and get the largest loan ever given by the Rural Elec-trification Administration. They award the first $150 million in contracts.

Now it's time to start acquiring the right-of-way. At this moment the farmers learn what is about to happen. They are outraged. The way they see it, everything has been kept secret until so much money has been committed that there can be no serious question of not building the line. At most, you might hope to push it over onto someone else's land.

But it so happens that this new plant will produce far more electricity than the customers of the two companies will be able to use for many years. It further happens that both companies have inverted rate structures – the more you use, the less you pay per kilowatt. And both actively promote what many consider to be inefficient uses of electricity, such as electric heating for houses in a cold climate. All this sets some farmers thinking. Maybe the line isn't needed. Maybe if the companies promoted conservation instead, it would never be needed.

Thoughts turn to action. In two of the counties the line is sched-uled to cross – Pope and Stearns – the farmers do what farmers seldom do. They organize. They decide to oppose the line – legally, of course. At that time, anything even faintly illegal would have shocked them. Many of these farmers are Norwegian-American, and incredibly law-abiding. Many have never had so much as a traffic ticket.

Very soon, they score a success. In Minnesota at that time, the right to use eminent domain was granted county by county, with state supervision. And in Pope County, the commissioners do something unheard of. They deny it.

But the law is just in the middle of changing. Power is to be centralized in the state Energy Agency. By a quick legal coup the power companies are able to transfer their request, and they shrug off the authority of Pope County. One down.

So the farmers hire a lawyer of their own and go to court. They are going to present two arguments: first, that the line is unnecessary, and second, that it may be dangerous. There are as-yet-unknown hazards in so huge a direct-current voltage. You could safely bury the big wires, of course, as the companies already plan to do when they near Minneapolis – but the whole line? Ridiculous, the companies say. It would cost seventeen times as much.

To the farmers' surprise and indignation, they are not allowed to present their arguments. Neither one is relevant, the judges say. What would be relevant? Well, if some farmer could show just cause why the line should not cross his land and therefore should cross some neighbor's land, that would be relevant. Appealed to, the state supreme court agrees.

By now, the people in Pope and Stearns counties are beginning to distrust the system. If they could have known what the deputy director of the Minnesota Pollution Control Agency would later say (just after resigning), they would have felt even more distrustful. He said he'd come to believe that his agency and the state Environmental Quality Council – both involved with the power line – were "set up to aid large corporations in getting around local government."

In June 1976, when surveyors for the companies began to lay out the line in Stearns County, they were met by a crowd of about sixty farmers blocking their path. The farmers were polite but determined. A little angry, too. Angry, for example, that when they had taken an appeal to the Environmental Quality Council, the hearing was presided over by a brand-new state official – an employee of the council whose previous job had been with the consulting firm in Michigan that sited the power line in the first place. They considered that suspicious.

Some of the farmers had their tractors with them; a few were on horseback. The surveyors retreated.

They were soon back, accompanied by reluctant local law officers. Over the next few weeks, the blocking party grew to several hundred men and women. There was little violence, just a lot of civil disobedience. Almost all of it occurred on land actually owned by one or another of the demonstrators. Many farmers had trouble believing that it was an indictable offense to demonstrate on their own land. A few people got arrested; a few went to jail.

These demonstrations spread from county to county in western Minnesota, and for a time the surveying almost stopped. Then the companies struck back. They started filing half-million-dollar damage suits against individual farmers. Arrests are one thing, but a suit that could cost you your farm is another. A lot of demonstrators withdrew.

But in Pope County, among other places, the resistance continued, and it began to get rougher. Survey stakes tended to vanish as soon as they were planted. By January 1978, the survey was still incomplete, though more than two hundred state troopers (out of a total state force of five hundred) were now patrolling the power line rather than the roads. That month, the Pope County attorney resigned rather than prosecute a group of newly arrested demonstrators.

Two hundred troopers can patrol a lot of power line, especially if they are helped by three hundred newly hired security guards. By summer, the survey was complete, and the 1,685 steel towers were fast going up. It was then that the Minnesota bolt weevils appeared.

The towers that carry a high-voltage line are somewhat like things you'd build from an erector set, only much, much bigger. They are held together with huge steel bolts. Unbolt the base of one, and at considerable risk to yourself you may be able to bring it down. On the night of August 2, 1978, the first tower crashed. Within a week, three more came plunging down.

No one has ever been convicted of sabotaging those towers, though it's not for lack of trying. The companies offered a reward of $50,000 for information, and later upped the price to $100,000. No takers. Instead the towers continued to crash. Over the next year, ten more came hurtling down, and many others were unsuccessfully assaulted. About midway in that year the companies gave technical ownership of the line to the federal government, so the FBI could come in. (The companies reassumed title in 1985. To the farmers, one more bit of legal sleight of hand.) The FBI caught no bolt weevils, either.

The farmers also didn't stop the line. The full force of the state of Minnesota was too much for them. The line was energized ten years ago, and it has been in use ever since. A couple of more towers toppled as lingering acts of defiance – the final one on the night of August 2, 1983, the fifth anniversary of the first. All has been peaceful since.

So what did all those criminal acts accomplish?

Well, quite a lot. For one thing, the companies had originally intended to build another seventy-eight-mile line, the so-called Wilmarth Extension. It remains unbuilt, and most people in Minnesota agree that violent resistance to the original line is the reason. More important, *any* new power line in Minnesota would now have to be justified in advance far more rigorously than this one was. The man who was governor of Minnesota at the time the towers crashed is on record as saying it is because of the farmers' actions that his state now has what he claims are the best energy-control laws in the nation.

More important still, the ripple effect has reached well beyond Minnesota. Not in public perception, because there hasn't been much of that, but in professional energy circles. It has reached Texas, for example. There, a similar 800,000-volt direct-current line has been killed by the Texas Public Utilities Commission. The Texas commissioners were swayed by a health study, done in Minnesota at the time the last towers were crashing down and undertaken because they were. That study was inconclusive – nowhere near proof enough to persuade Minnesota to take the line down again. But it raised enough doubts, particularly about ionized air, to stop the new one in Texas.

And then there is a final effect that is highly ironic. One of the arguments the Minnesota farmers wanted to make in court was that a high-voltage DC line may have special dangers – may be worse for people and animals living within its magnetic field than the more usual alternating current lines. Silenced in court, they resorted to ecosabotage; and one consequence of their expensive ($6 million in damage), destructive, risky, illegal action was to encourage a more careful study of power lines in general than had hitherto been done in this country. As a very indirect and indeed unprovable effect, research got funded that might not have otherwise.

It now begins to look as if DC lines are actually safer, though as the Texas commissioners point out, this is far from proven. But don't relax: what that means is merely that high-voltage AC lines may be more dangerous than anyone has realized. A new study made in Denver suggests that up to 15 percent of the cases of childhood cancer in this country may be due to the electric and magnetic fields of high-voltage AC lines. If *that* is true, then the seventeen-times greater expense of burying power lines (if that power-company figure is correct) will come to look remarkably cheap.

Were the bolt weevils justified? To me, it seems they were. If a new power line equally dubious in origin were to come across my own farm, would I join a similar group? I don't know. I do know that I'd be tempted. And considering how much I dislike violence, how deeply I believe in a government of laws, not to mention how afraid I am of even traffic tickets, that's saying a lot.

[1989]

The Soul of New England

🎐

RECENTLY I LEFT my farm for a week. I flew from Vermont down to Boston in a little plane and then from Boston out to California in a big plane. I was going to a conference at that exurb-of-Los-Angeles branch of the University of California called Riverside. And never having been in southern California before, I had come a day early. I wanted to see the landscape. I wanted to do it the only way worth doing: on foot.

As it turned out, the landscape made me nervous. There were plenty of individual elements I loved, such as the double rank of bottle brush trees marching up the avenue at the main entrance to UC Riverside. But the countryside as a whole was too naked for my taste. All those bare, lion-colored hills and all that sweep of sky left me feeling exposed and vulnerable – a little bit like how a mouse would feel (I imagine) if you put it out in the middle of a basketball court. A New England mouse, anyway.

The evening of that first day I devoted to looking at the city of Riverside itself. It's a small city by California standards but quite a big one by the standards of New England. It has 226,000 inhabitants, which makes it more than twice as big as anything in Vermont, New Hampshire, *or* Maine. In all New England, only Boston exceeds it.

I was staying in a motel out by the university, and what I did was to stroll downtown just before dark. Some stroll! In the first place, downtown proved to be elusive. I'm not entirely convinced

Riverside has one, at least not in the eastern sense of a thronged center where building density is much greater than elsewhere, where you can't easily park, where things are *old.* In the second place I remained tense. It was the same mouse feeling I'd had out on the bare hills. Only this time the streets were the cause.

I was walking down a broad avenue which crossed side streets at regular intervals. Every time this happened, I went into a state of mild panic – that is, I panicked once a block.

Don't scorn me entirely. Those were formidable side streets, like nothing I had ever seen before. Each had six traffic lanes and two parking lanes, for a total of eight. At home in Vermont not even the two interstate highways are that wide. I'm not sure the Connecticut River is.

I had to force myself to cross. After the first couple I no longer had any fear of being squashed out in the middle by some crazy California driver. California drivers proved not to be crazy, instead quite gentle. No one shot round a corner with squealing tires or threatened me by speeding up slightly, as drivers sometimes do in Boston. And as they even do in West Lebanon, New Hampshire, for that matter, in front of the shopping malls.

My fear – I knew it at the time – was quite irrational. Simply, I felt too far from the nearest tree. Or, rather, this being a city, too far from the nearest doorway. Even when I had made it across a street, I was still a long way from shelter. These eight-laners were not bordered by comforting buildings. They were bordered by emptiness. Every laundromat and photo shop had been set back at least a hundred feet, and the intervening space was generally paved. I couldn't even have burrowed into the soil, had a helicopter come thumping down to catch an unwary visitor.

That trip taught me a lot about California, but it taught me even more about New England. Opposites may or may not attract; they do clarify each other. And northern New England and southern California must be about as opposite as any two parts of the United States can get.

The Soul of New England

They have things in common, of course, both being part of American culture. Teenagers in East Topsham, Vermont, listen to the same American Top Forty songs as teenagers in Redondo Beach, and (except all winter) they wear the same gaudy sneakers. They don't eat the same too-rapid food as much, but that's only because they have to go all the way into Barre to get it. Fast food is not available in Topsham. Church suppers are.

So it goes through the whole spectrum, from theology to garbage collection. There are Hindu ashrams and Buddhist monasteries in New England as well as in Los Angeles. Compactor trucks in both places make the same unearthly whine as they swallow the same plastic garbage. Even scenically the two regions share much, from identical phone poles to interchangeable road signs to similar-looking storage buildings slapped together of cement blocks and then painted pale green. (Somehow I find them even uglier in New England than in California, probably because I have a clearer image in my mind of what a New England building ought to look like.)

But if there are many likenesses, there are even more opposi-tions. As I came back on the jumbo jet to Boston and then on the little plane to Vermont (nineteen passengers, no cabin attendants), I kept thinking about them. It seemed to me that I saw the true nature of New England more clearly than ever before.

Let me start with the landscape, then advance briefly to the people and finally I'll say a cautious word about the soul of this region. Not all regions have souls, at least not living ones, but New England does.

The central truth about our landscape is that it's introverted. It's curled and coiled and full of turns and corners. Not open, not public; private and reserved. Most of the best views are little and hidden. It was only after I started doing contract mowing of hayfields around town that I got behind people's houses and saw vista after vista that you'd never guess from the public roads. We like secrets.

Did I call our roads public? They mostly aren't, other than in the legal sense. They are narrow, except for a few monsters curving around Boston and Hartford. Even when country roads are bordered by fields, there's apt to be a line of trees and brush that effectively secludes the traveler. Or if they do offer a view, it's small and domestic. Even a New England interstate can offer private views. In fact, the road I'd pick to be emblematic of the region would not be one of the thousands of little dirt roads curling and coiling all over the six states, including Rhode Island. They are indeed emblematic, but perhaps more of the past than of the future.

What I would pick would be Interstate 91, the stretch beginning at mile 112 in Vermont, going north and coming over the crest of the high ridge just beyond Wells River. That section doesn't look like an interstate at all. What you see in the valley below is a pastoral scene with what appear to be two separate two-lane roads running parallel, a hundred yards apart. What is actually a small rest area gives the impression of being a side road, curving off to the right. In short, the whole road project has been domesticated, and instead of cutting the land apart it stitches things together. And that is an emblem of how our humanmade landscape, at its best, interacts with what was originally here. I greatly prefer it to big faces on Mt. Rushmore, or the six- and eight-lane concrete strips that press down so hard on the poor suffering but still beautiful meadowlands of northern New Jersey.

Let me take my own farm as an example of how the land is arranged. If you were to drive by it on the town road, you'd notice a fairly handsome old brick house, with a big red double barn attached to the western end. Across the road you'd see a few acres of cow pasture, fenced partly with barbed wire and partly with very handsome stone walls. (You'd better think them handsome. I built them.)

That's all you'd see, but it's hardly all there is. Behind the house, completely invisible except to one set of neighbors across the valley, is my best hayfield – which is also the best kids' sledding. I almost lost a friend who couldn't resist observing, every time he came to visit, what a fine three-hole golf course that field would make.

The Soul of New England

Across the road there is a great deal more. Behind the field you can see from the road, three other fields curl around a steep hill, and a steep pasture goes up it. To walk from one field to the next – *this* with its pine-covered knoll where the cows like to hide when they are about to calve, *that* with its oak tree older than the place, and its high cliff boundary – is like walking from one stage set to the next. Everything is so up and down, and so deeply wooded where it isn't pasture, that my little ninety acres offers a dozen places to get lost. Even now I don't know every inch. It was only six years ago – nineteen years after I bought the place – that I stumbled on the fort. It's a giant rock that some glacier left: the length of a school bus, but higher and wider. Centuries ago frost split it lengthwise, and then split one of the side pieces as well, so that there is a T-shaped passage right through the great rock. A hundred-year-old yellow birch grows in that passage; dense woods surround the entire rock and conceal it.

What caught my eye as I pushed past the last bull pine was the fortification. At some point long ago children rocked up each of the three entrances to about waist height, and at least one of those children really knew how to lay up stone. It is beautiful work. An average of one new person a year sees it. I have begun to worry that I am showing it off too much.

That rock is not the sacred place on my farm; the huge old oak marks that. But it may yet become sacred – in fact, the whole farm seems to be in the process. Which is one reason why I have taken steps to keep it a farm.

There are sacred places on nearly every piece of land in town, except maybe our one small commercial zone. And I would claim that the deepest truth about New England as a place is that, with the exception of some Indian reservations, it contains a higher proportion of sacred land than any other part of the United States. By "sacred" I obviously don't mean formally consecrated to a religious purpose – though there's a fair amount of that, too, around convents and monasteries in Massachusetts, not to mention around the lamasery thirty miles north of here. I simply mean land valued other than commercially – land for which the highest use (as tax appraisers quaintly say) is not discovered by finding what will bring the biggest cash return, but by finding what will make the land most beautiful, most productive, or most healthy, and sometimes all three together. And, yes, when I make that claim, I speak in full awareness of the million acres of the Adirondack Preserve in New York, and the Amish country in Pennsylvania, and all those Civil War battlefields in Virginia, and Yosemite, and....

I'll offer just one piece of evidence for my claim – but what a piece! Nine years ago, when I was first looking for ways to protect my farm from high uses, I did some investigation of private land trusts: organizations dedicated to land preservation. At that moment there were somewhat under five hundred of them scattered across the United States. Some states had one or two; some had none. Connecticut had eighty-two and Massachusetts sixty. No state outside New England came even close to these figures.

The people of New England are a good deal harder to generalize about than the land. We are a seacoast people and a lakeshore people, as well as land-lovers. There are sacred coves, too. My wife grew up on one, and even now she wishes it were possible to hear foghorns on our farm. We may not have any seriously large cities except Boston, but we have hundreds of mill towns, some grimy, some not. We are partly French-Canadian, partly old Yankee, partly Italian, partly twenty other things. There are Mashpee Indians on Cape Cod, and there is a jai alai fronton in Hartford. There is now one county in Vermont that has an actual majority of upscale newcomers, and in that county a term like "Sunday brunch" is heard more frequently than a term like "church supper." I believe they have hot tubs, too.

But climate and topography do play a role in determining human character. And the example set by native Vermonters does seem to have an effect. That same county also has an enormous number of houses heated by woodstoves, and the sale of chainsaws to newcomers is brisk. Despite the vast changes of the last twenty years, I think it is still accurate to say that the basic New England characteristic is a kind of humorous stoicism. You *expect* it to snow just before you have to drive a hundred miles, and to be sleeting when you have a day off to ski. You are not surprised when your pipes freeze, and you probably have a wry comment to make. I love one I heard last fall, made by a woman in New Hampshire. She herself lives in a small city (about one-eighteenth the size of Riverside), but she has a daughter in the country who runs a cider press. She had filled her car with apples at that press, to take to another daughter in town, who was going to make applesauce. Before she could deliver them, a storm came up, and blew over a fifty-foot tree in her yard, on top of car and apples both. There are a good many things one might think to say at such a moment. "It sure put a cleat in my car," she said. "I guess I've got a convertible now." Sounds just like New England to me.

What is the soul of New England? Something inward, something a little cold even, at least that's how it's going to strike a newcomer. But something fiercely determined, and even more fiercely protective. Almost relishing discomfort. Able to endure almost any adversity, and just get stronger. The one thing that may sicken it is too much ease and prosperity – which, indeed, I suspect is true for almost every region with a soul. Somewhat more tied to the past than any other part of America except a little bit of the South. It has a longer past *to* be tied to than any other part of America except a little bit of the South.

And yet any living soul can change, and must. New ideas gather in New England with some frequency. Once it was the idea that slavery should be abolished. Then the idea that everybody should be educated. Right now two ideas are strong. And hence two changes I have seen in my own conservative Vermont over the past twenty years are first a mighty tide of environmentalism and second a very rapid alteration in the relation between the sexes.

One of the events I missed during my week at Riverside was a barn-raising on a new organic farm on the other side of town. When I got back, an aged neighbor (still able to use a hammer) was telling me about it. He had been there, helping. "You ain't going to believe this," he said gleefully. "They was twenty carpenters up on that roof – and damn near half of them was *wimmen!*"

There are barn-raisings still to come in New England. New Englanders of every stripe will be up there, hammering and talking, preserving the sense of community that has kept us going for the past three hundred years.

[1989]

My Farm Is Safe Forever

🝰

EVERY DAY THERE ARE ninety-three fewer farms in America than there were the day before. Some get amalgamated into agribusiness holdings, and a few are simply abandoned. Most, however, get paved, built on, developed, or occasionally turned into nature preserves.

None of these fates awaits my farm. It's going to stay a farm long after I have moved into the village cemetery. Long after my grandchildren – and I don't even have any yet – have done the same. In fact, forever. And if that seems too huge a claim, just wait a minute.

I can speak with such confidence for an excellent reason. A few years ago I made a solemn and binding agreement with the small Vermont town I live in. I gave the town the development rights to my farm. For its part, the town agreed never to use them. If the town ever changes its mind – if, say a hundred years from now, whoever is running things gets tired of holding the rights – they automatically pass to a private conservation group three towns away. Should that no longer be around, the development rights go to its successor organization.

Meanwhile, I still own the place. I still possess every right of ownership that I care about. I can continue to raise beef cattle, bale hay, make maple syrup, cut logs, make whatever rural use of my ninety acres I feel like. If that begins to bore me (it won't), I can sell to any buyer I please. I could even sell the place to McDonald's.

It's just that if they bought it, they couldn't put up any golden arches. They'd have to install a farm manager, and start raising beef cattle, making maple syrup, or whatever.

I gave up the right to develop my farm solely to protect the land, but as a kind of bonus there are some pleasant cash benefits. Getting rid of the development rights has assured lower taxes on the place, for me and for all future owners.

The taxes will stay exactly the same on the house and on the two acres around it – what our state government calls a homestead. But on the other eighty-eight there will be a highly pleasing difference. From now on, these eighty-eight acres – my present fields and woods – will be taxed on their value as farmland and woodlot, not on what they might be worth if they were converted into forty-four building lots, or a tennis club. (That forty-four is not random. It's the maximum number of lots a developer could impose on my farm: we have two-acre rural zoning in this part of town. Actually, what with the roads the developer would have to put in and the steepness of a couple of my hills, I suspect he or she would have to scramble to put in more than about thirty-five new houses. That's still thirty-five more than I want.)

Back to the money. As I was saying, the taxes are now lower. But that's not all. The difference between my place's value as a farm and its value as developable land, which the town listers worked out to be $27,800, counted as a charitable deduction on my income tax. Need I say I've never had a deduction like that before? Since my income was nowhere near high enough to use it all in one year (still isn't), I had a carryover deduction for the next year – and there was still some left for the year after that. Most fun I've ever had with the IRS, or ever expect to have.

This was no slick private deal, either. It was not arranged or suggested by a tax lawyer. It is something almost any farmer can do. In fact, you don't even have to be a farmer. You just have to be the owner of a fair-sized piece of open land that qualifies under "a clearly defined federal, state or local government conservation policy." I'm quoting IRS Publication 526, entitled "Charitable

Contributions," the same lovely document that enables me to talk so grandly about permanence. The tax benefits apply, says Publication 526, only if the land is "protected forever." Suits me. Even though I secretly know that "forever" means something like "as long as the United States exists in its present form." That's likely to be a while.

The luck is that such laws exist. It took more than luck, however, to get my farm protected under them. It took persistence, a good bit of paperwork – and, of course, the willingness to give up the profits I might have made by selling to a developer.

I first began thinking about protecting the farm in the spring of 1980. Two things set me going. One was the particularly brutal fate of a farm a few miles away, a dairy farm that had been in the same family for more than a hundred years. That place wasn't developed; it was dismembered. It had been my idea of a perfect farm.

The other was that my children were growing up. I have two, both girls. The older, having lived in rural Vermont all her life, was talking longingly about London. The younger was beginning to dream of being an airline flight attendant. It seemed clear to them that they didn't want to be farmers or farmers' wives, or country dwellers at all. It seemed clear to me that when I died the natural thing would be for them to sell the place. Having spent twenty years restoring it to a beautiful and moderately productive farm, I didn't relish the idea of bulldozers leveling my carefully rebuilt stone walls or blacktoppers advancing into the orchard.

The first step was a serious discussion with my daughters. I began with a little lecture on what it means to own a piece of land. Not what it means emotionally (they've known all about that since they were little), but what it means legally. To own a piece of land, I explained to two moderately bored teenagers, is to own a bundle of rights, most of which can be separated from each other. For example, you can detach the mineral rights from a piece of land and sell them, while still keeping the land itself. You can also detach the right to develop, and lock it safely away. That was what I wanted to do. If I did, it would mean my daughters would inherit something considerably reduced in value. Reduced $27,800, to be precise, though of course I didn't have that figure then. It would be even more now.

In some ways Americans are the least materialistic of all people. The girls didn't hesitate a second. "Dad, we know how you love the place," my elder daughter said. "We love it, too. We just don't want to spend our lives looking after cows. You go right ahead."

Despite their encouragement, I didn't do anything more for about a year and a half. Partly that was to give the girls a chance to change their minds. Mostly it's just the way I operate. Slowly. I'm the sort of person who can decide he needs a tractor and then spend several years thinking about it before actually going out and buying one. During the next eighteen months I collected a fat folder of clippings about development rights, but that was all I did.

Then in the fall of 1981 I finally made a move – in fact, a double move. We have a planning commission in our town, and I wrote its chairperson to ask if the town would be interested in a gift of development rights. And I put the same question to what was then the Ottauquechee Regional Land Trust and has since tripled in size and become the Vermont Land Trust. At the time, it was the nearest to me of the many hundreds of land trusts that have sprung up across America in the last ten or fifteen years. (There are also some, mostly in the Northeast, that have been around much longer.) I knew about Ottauquechee from my clipping file.

Not surprisingly, the planning commission replied cautiously, since no one had ever made such an offer to the town before. They said come to a meeting and tell them about it, which I did. Then they deliberated for three months. Then they decided to pass my query on to the selectmen. (For those who don't follow the minutiæ of New England customs, selectmen are what we have as local government, rather than mayors and councils, county commissioners, and the like. Some towns have five selectmen, some three. This one has three.)

Also not surprisingly, the Ottauquechee Land Trust replied with a good deal more zest. Though only four years old at the time, it had lots of experience with development rights, and at that moment was working on roughly a dozen cases. Richard Carbin, the vigorous young director, first wrote me, then phoned, then came over and walked the farm with me. That was to make sure it really could function agriculturally into the indefinite future. On a small scale, it can. I've got two or three nice pastures. Any time I was ready to protect them, he said, the trust was ready to cooperate.

Meanwhile, the selectmen had also been thinking. Their decision was that something like this had better come up at Town Meeting. If the voters approved, they wouldn't oppose the gift. If the voters turned it down, naturally that ended the matter.

Furthermore, instead of placing my offer on the agenda themselves, the selectmen felt I should do it by petition. That's work. Such a petition must be signed by 5 percent of the voters in town, which at that time was sixty-five people. (It's currently seventy-four.) Getting up a petition is not difficult or technical – every year we have two or three articles to discuss at Town Meeting that have been petitioned by individuals rather than warned by the selectmen – but it does take a little time and effort.

By now we were well into 1982. I had a choice to make. I could quickly donate rights to the land trust. Or more slowly, and with some bother, I could give them to the town.

I chose to go the town route, and I had three reasons. One was emotional. I like the direct democracy of Town Meeting, and I liked the idea of the voters' making the decision. One of the ways we educate ourselves in rural New England is by arguing things out at Town Meeting.

The second reason was practical and involved taxes. The big question for both the planning board and the selectmen had been precisely how much real estate tax the town would lose if my farm ceased to be developable. Towns hate losing revenue. The question was complicated, because my thirty-two acres of pasture, though not the fifty-six acres of woodland, were already enrolled in a state program of current-use taxation. The town taxed part of the place as farmland, the rest as potential development land, and the state made up the difference. All of us assumed the state contribution would cease once I gave away the development rights. Our best guess was that the town would lose about $400 a year if the gift went through.

Of course I could have thumbed my nose at the selectmen and gone straight to the land trust – but only at the cost of enraging some of my neighbors and maybe giving conservation a bad name locally for years to come. If my taxes went down, wouldn't everybody else's have to go up, however slightly? Wouldn't it in fact look like a slick private deal? I preferred to face people at Town Meeting and so get a chance to make the case that in the long run land protection holds down everybody's taxes.

That's where the clipping file would come in handy. I thought the story of Suffolk County, Long Island, for example, might catch people's attention. I had a clipping that said the county authorities there had raised a sum of $60 million to *buy* development rights on farmland. They figured on protecting twelve thousand acres at a cost of $5,000 an acre. And then they expected to recoup that whole enormous sum "in terms of tax dollars that won't have to be spent on more schools, roads, and services." If purchased rights pay for themselves, what a bargain free ones must be.

I had another good clipping, too – this one about a county in Virginia that did a close study of where it got its money and where it then spent it. Turned out that for every tax dollar the county took in from farmland and woodland, it only had to spend about seventy cents. But for every dollar it got from developed land, it spent $1.11. It made sense to me. Cows don't go to school. And counties don't have to maintain farm roads.

Finally, I had a personal financial reason for donating my development rights to the town rather than the land trust. Land trusts need revenue just as much as towns do. They've got to have offices, phones, files, legal advice, the lot. The older ones have depended on the traditional income sources of nonprofit organizations: membership dues and fund drives. Many also keep their expenses down by hiring no paid staff whatsoever. In Connecticut, a pioneering state in land conservation, there are eighty-two small land trusts, almost double the number in any other state. According to Allan Spader, director of the recently formed Land Trust Exchange in Boston, all but two or three depend entirely on volunteer staff.

But some of the newer land trusts, with much to do and little time to do it in, are experimenting with a sort of fee-for-service approach. The donor of development rights is asked also to make a cash contribution which will become endowment for the trust. Montana Land Alliance uses that approach and so, as it happens, does the Vermont Land Trust. It asks for 3 percent of the value of the land to be protected.

That seems perfectly reasonable. We who donate the land get tax savings much larger than 3 percent of the value of the land. The trust has an urgent need for money. Fund drives are more difficult to conduct in a poor state like Vermont than in a rich state like Connecticut. All that I see. But I have the misfortune to be a born tightwad. If I could save the 3 percent, I still preferred to. I began to prepare for the 1983 Town Meeting.

Getting the signatures for my petition proved to be extremely easy. I had been dreading it, because I hate going house to house asking for something, even if it's only signatures. Then the woman who runs the store in our village made a good suggestion. I wrote a brief account of what I wanted to do, stapled it to a sheet of blank paper, and left it on the store counter.

Within a week I had almost fifty signatures. When a second week produced only three more, I made a copy of my statement, and with the owner's consent left that on the counter of the general store in one of the other villages in town. Quite soon I was up to sixty-seven signatures, two more than needed. When the warning for Town Meeting was posted on January 24, 1983, Article XI read, "To see if the Town wishes to accept a gift of the development rights to Noel Perrin's farm in Thetford Center."

Before Town Meeting we have an event called Pre-Town Meeting. It comes a week ahead. All the candidates for office attend; so do people who have a petitioned article, and usually about fifty or so of the more serious voters. There is time to go into issues more deeply than is sometimes possible at Town Meeting itself.

My proposal got a cool reception. No one actually said "slick private deal," but clearly the people who asked me questions were thinking about the lost $400. Someone mused out loud on what would happen if other landowners began doing the same thing – what would happen to his own taxes, that is. Naturally I had my clipping with me. But he and the other questioners were not much impressed by the news from Suffolk County, Long Island, or even by what was happening much nearer at hand in Massachusetts. *That's them; we're us,* seemed to be the feeling.

I went home depressed. It didn't cheer me a bit when the chairman of our Board of Selectmen (I've known her for twenty years) called me that evening. "I'm going to move to pass over your article," Ginny said. "You'll do better to wait a year. It doesn't stand a chance now."

People who work for land trusts are generally idealists. I had stayed in touch with Rick Carbin at Ottauquechee all *the* time. (I had also become a dues-paying member of the trust. Still am.) He wasn't in the least touchy that I was going the town route; what mattered was protecting another farm.

The next morning I phoned him for advice. Don't even consider giving up, he said. Instead, quickly get in touch with the town clerk of Pomfret, Vermont. She can tell you something that will interest the voters in your town.

The result was that when I went to Town Meeting, I not only had with me a list of facts and figures, such as that we had lost 153 farms in Orange County, Vermont, in the preceding ten years, I had a brief letter from Hazel Harrington, the town clerk of Pomfret. That was my secret weapon. Only the head selectman knew about it. It had persuaded her not to pass over Article XI, after all.

When it came my turn, I made a brief but passionate speech. Just to warm up, I mentioned Long Island, Massachusetts, and the lost 153 farms in our own county. Then I brought out the secret weapon. I told the meeting there was one precedent for what I proposed to do. In 1981, Henry Bourne of Pomfret had deeded development rights on a piece of land to his town, the first person in the state to make that kind of gift. Then I read Mrs. Harrington's letter.

Dear Mr. Perrin,
The land on which Mr. Bourne deeded development rights to the Town of Pomfret is listed no different for tax purposes than it was before he took this action. He retained title to the property. However, he as an individual is in the current-use program. Therefore, the Town is getting the same amount of taxes on this property as before.

Thetford Town Meeting took that in. Then it heard me promise to get my woodlot into the current-use program, along with the pastures. Then a couple of people asked perfunctory questions. Then the town voted overwhelmingly to accept my offer, and we went on to Article XII.

And that's how my farm comes to be safe, if not forever, at least for a long time to come. That's how I came to have a $27,800 income-tax deduction for doing what I wanted to do anyway. If you are a landowner, you, too, can probably make money by protecting your land. Ecologically, aesthetically, and morally, that seems preferable to the traditional practice of making money by destroying it.

[1984]

Postscript, 1991: It's now eight years since that town meeting. Thetford is under a lot more development pressure than it was in 1983. It is also much more used to the idea that when developers push one way, it is perfectly all right to push back the other way. We now have a town Conservation Commission, which played a key role in saving Earl LaMountain's farm last year. That's a full working farm, as opposed to my half-working farm.

Its members were also busy two years ago, when the old Shyott place got sold to a developer. The sale was not stopped, nor am I sure it should have been. There are three million more Americans every year, all needing a place to live. Lots of them want to live in the country, and Thetford will get its share, as it got me, thirty years ago. The good news is that part of the Shyott place will stay country for them to live in. The big twelve-acre hayfield at the front will stay a hayfield, and the handful of new houses will be up behind, well spread out. Under pressure, the developers themselves put on conservation restrictions.

The Conservation Commission didn't bring that about all by itself. We also have a new and even closer-to-hand land trust. I belong to that one, too. The Upper Valley Land Trust was the other key player in saving LaMountain's farm and keeping a rural aspect to Shyott.

What about my own farm? That hasn't changed a bit, except that I finally ditched the upper meadow, and I've put up one new stone wall. My older daughter did get to London but came home after two years. She now lives in Seattle. The younger one is in Boston. Taxes continue to climb, of course. It's no longer $400 a year that I save by having conservation restrictions on the place; last year it was $817.03. The town got reimbursed by the state, same as always.

What if one day the state gives up its current-use program, and therefore quits reimbursing towns? That would certainly be a blow. A minor blow. The taxes of everyone in town, me included, would go up an average of sixty cents.

On the other hand, suppose I had sold out eight years ago. Suppose there were now thirty-five houses on what used to be my farm. In *that* case, everybody's taxes would have gone up somewhere between $100 and $500. I still think the town got a bargain. And so did I.

Uncollected Pieces

A House, a Horse, a Hill, and a Husband

🕊·🐿

I MET ANNE in the summer of 1988. We were both teaching at a writers' conference at Manhattanville College, forty miles north of New York City. She had the class in writing for children, I the class in writing nonfiction.

Anne was someone you noticed right away. She was very blond, very blue-eyed, astonishingly young looking for someone who was rumored to have grown children. Rumor also said she was newly single and planned to stay that way. Pity. I, too, was single with grown children – only I'd been single for eight years and was ready for a change. For one thing, I wanted someone to share my beautiful old brick house with me.

You'd think that at a small writers' conference, where there are only five teachers and all five are assigned rooms in the college guesthouse (a building not much bigger than my farmhouse in Vermont), we'd all get to know each other in a hurry. Well, four of us did.

Anne, however, proved inaccessible. During the days of that busy week, we were *all* fairly inaccessible, except to our students. We taught classes in the morning and held individual conferences in the afternoon. Anne held the most. Or at least she held the most outdoor ones. Often I'd see her blond head under a tree, where she'd be sitting with a would-be children's author. The afternoon would get late, a different student would be sitting with her, but the conferences went on.

Then, just as free time began, she'd vanish. Of course she had a room in the guesthouse, but she never slept there. I never even saw her drive away. She just disappeared every evening, like a princess in a fairy tale.

Was this part of her stay-single program? Could be. If you avoid meeting people, if on social occasions you are simply not there, you are going to be fairly safe from potential husbands.

Or was she actually slipping off to a glamorous other life, something beyond our ken? This was the view we other four inclined to. We had no proof whatsoever, but we did have some strong clues. We had known all week that Anne's last name was Lindbergh, and that, yes, she was the daughter of Charles and Anne Morrow Lindbergh. It is one of the civilized customs at writers' conferences to read at least one book by each of your fellow teachers. So we also knew that she was no secondhand celebrity, not merely the beautiful daughter of famous parents. She was a gifted writer. That week I had read her *Hunky-Dory Diary* and *Bailey's Window* and loved them both. Anne gave her characters second chances, which is what life should do, too.

We even knew some personal details, such as her age. She was forty-seven. (One of our two adult novelists figured this out from one of Anne Morrow Lindbergh's published diaries. Birth of daughter, October 2, 1940.)

Finally, we knew that Manhattanville was an easy train ride into New York. Stands to reason that she was vanishing into the city, where elegant friends awaited her.

But the last night of the conference Anne didn't vanish. She slept in the guesthouse. Almost had to. There was a long evening program, and she couldn't have slipped off to wherever it was she went much earlier than Cinderella dashed out of the palace.

I keep Vermont hours. The final morning I was up and out by 6:30, an hour at which the woods behind the Manhattanville campus are cool and fresh, even in late June. The first (and only) person I saw when I stepped out the door was Anne. She obviously was planning a walk, too.

"Have you seen the ruined chapel?" I asked. She hadn't.

"Want to walk down there with me?" She did.

"Where do you go every night?" I asked after a while. "We're all dying of curiosity."

She laughed. "You're going to be disappointed. My mother lives in Darien. She's been sick. I've been staying with her."

"So where do *you* live?" I asked presently.

She laughed again. "The same state you do." Ha! Clearly she, too, had been reading her colleagues' books.

"You live in Vermont?" I said, just to be sure I'd heard right.

"Yes. I have a little farm near St. Johnsbury – bought it last winter. It's nothing grand. Just a house, a horse, and a hill."

The gods were smiling. St. Johnsbury is less than fifty miles from my farm.

By now we had reached the ruined chapel, deep in the woods. To get into it, we had to climb through one of the ruined windows. I went first, then offered her what was doubtless a superfluous hand. She hesitated, then took it. Before breakfast, she had agreed to come down some day soon and see my house, my hill, and my animals. (No horse. Just three sheep and eight cows.)

A few days later she drove down for the promised visit. She was wearing jeans and a blue workman's shirt. She looked ravishing.

"Let's save the house for later," I said. "I want to walk you around the farm, and then I thought we might go swimming in the river."

"I have my bathing suit on under my clothes," she said happily.

Anne loved my farm, as well she might. I have ninety acres of woods and pastures, lots of old stone walls plus two new ones, a hilltop with a 360-degree view. The house, when we came in for lunch, seemed to excite her less. She duly noted what a beautiful color the old bricks were – the house was built in 1820. She liked its rustic version of the Federal style. But something was on her mind.

"How long have you lived here?" she asked.

"Twenty-five years."

"So this is where you and your first wife lived?"

"Yes."

"And your second wife lived here, too?"

"Part of the time," I answered guardedly, since it was clear that these ghosts of former wives did not appeal to her.

I can't say this surprised me. My second wife had not at all wanted to move into the house, either. She would rather have started fresh. But I had so much work and love invested in the farm itself that I couldn't bear to leave. One whole hillside was covered with trees that I had planted as seedlings; the hay barn had taken several hundred hours to repair and repaint; the new stone walls were ones I had built myself.

In the end, my second wife and I had made a deal: Give her a free hand with renovations, and she would agree to live in the house. The renovations were so extensive that we didn't move in until a year after our marriage, which is why I could answer "part of the time."

Anne said nothing more about the house that day, except to comment that the living room (as redecorated) reminded her strongly of a candy box. And she may have said something about the scarcity of flowers in the yard, I having neglected the garden during the eight years since my second divorce. I was too busy with walls and cows and hay.

We finished lunch. "I should go home. You've probably got things to do," Anne said.

But was there a trace of reluctance? I thought there was. "Not quite yet," I pleaded. "There's a birch wood I want to show you. It's just up the road."

"I've got all afternoon," she said simply.

When we reached the wood and she found there were many mushrooms growing along one edge, we *took* all afternoon – and then I persuaded her to stay for a church supper a few miles away in Lyme, New Hampshire.

A House, a Horse, a Hill, and a Husband

"Now it's your turn," she said, as she finally got into her car. "Come see my horse and my hill."

I would have gone the next day, if I hadn't been afraid of scaring her. So I went the day after.

But she was no more afraid than I. We got engaged with astonishing rapidity, and by early fall we were discussing where we would live. Meanwhile, I had of course been up to her farm many times, and it was clear to me that I didn't want to live there. Quite apart from loving my brick house, I had as many as three or four objections to her place. If I lived there, I'd have a sixty-mile commute to work. Her little farm, only eleven acres, was all on a north slope, and the soil was miserable. The actual house was the modern kind, all wood and windows, with almost no partitions on the ground floor. I like doors.

I did agree that her house, having all those huge windows, was much lighter than mine. I also liked its informality. But I minded that the sun vanished so early behind her hill – in the winter, about 1:30 P.M. Finally, her little bit of woods was an unholy mess of tenth-growth pines, nearly all crooked. My woods were big and beautiful.

And Anne had plenty of problems with my place, besides the ghosts of former wives, the too-small windows, etc. Her house was the first place she'd picked and owned independently, and she loved the sense of freedom it gave her. She didn't like the considerable and steadily growing traffic on my road – a paved road at that, no fun for riding horseback. Her own house was on a dirt road so remote and hard to find that she didn't even bother giving me directions the first time I came up, but just met me in St. Johnsbury and guided me out. If I would be quadrupling my commute by moving to what we were now calling North Farm, she would be increasing hers just as much by moving to my place, alias South Farm. She'd be increasing the distance from her sister, too, who lived just a couple of miles from North Farm.

And greater and more unanswerable than any of these, her

children were *not* all grown. The youngest was only eleven. Marek had taken our engagement hard. He had no wish to live in my house, or even to visit it.

Neither of us was prepared to give a flat no. If necessary, we would have set up housekeeping in a rain barrel in order to be together. During the fall, and before we got married at Christmas, we each set about making changes at the other's house, just in case. I replaced a particularly hideous wire fence at her place with a small neat stone wall, mostly bringing the stones up from South Farm, where there are more flat ones. She picked and I bought a second horse to keep at her house, so we could ride together. Anne de-candy-boxed my living room. (She was right; it had looked like one.) She began educating me about flowers. As she wrote a cousin who had sent her a huge bouquet for her forty-eighth birthday that October, "Ned was here for the weekend, and I got in a lot of comments about how wonderful your flowers were, and how much I loved flowers, and how seldom I get flowers, and other guilt-making, inspiring things."

But along about November we got our good idea – or, rather, the first of our two good ideas. Why should either of us have to move? It was forty-six miles, door to door. We would simply live alternately at North Farm and South Farm. We were the more willing because each of us had two failed marriages, and maybe the inevitable days and nights apart in this third marriage, to be lived in two houses, would help us stay forever in the glorious romantic haze we now occupied.

Certainly the fall went well. Our being sometimes apart lent itself to such incidents as the night in November when I couldn't go up and she couldn't come down, so naturally I phoned her that evening.

"You interrupted me," she said reproachfully. "I was writing you a letter."

So we got married – in her house – and for three years we lived a two-house marriage. Then Anne got the second good idea. By now we had both changed our minds about part-time marriage.

A House, a Horse, a Hill, and a Husband

We weren't yet at the point we reached later when we realized we would gladly spend eight days a week together, if it were possible to do this in a seven-day week. But the romantic haze was clearing, and what it revealed was a sort of spiritual mountaintop on which we stood and from which we could see a view of marriage more splendid than either of us had ever supposed to be possible.

Anne's new good idea was this: What we needed was three houses. The third one would be in the middle, between our places, and we would build it ourselves. At least initially it would be very small: one room, 16 feet by 18, with a sleeping loft. Later we'd add two small writing rooms, one coming out as a wing at each end. As befits the house of a woman so gifted at disappearing, it would be on no maps, have no phone, be inaccessible to people who are fascinated by writers or by Lindberghs. We could go there lots.

Marek was now away at school (and also friends with me). Any nights I couldn't get to North Farm or she to South Farm, we would meet at what we inevitably started calling Middle Farm.

I think Anne had one more reason to want a Middle Farm – but she was too smart to tell me this last reason, and I was too smart to tell her I guessed it. By now she knew that it would be almost impossible for me to decide by a sheer act of will to sell South Farm, which I had owned for twenty-eight years. (And almost as hard for her to give up North Farm.)

But if we together bought a piece of land and together built a house, my loyalties would gradually shift without my even noticing, and one day I would realize I could let the brick house go and not even care.

It worked. In 1991 we together bought a piece of land compared to which her eleven acres near St. Johnsbury and my ninety acres near White River Junction are practically urban. It is on no town road. Zero cars a day go by. Long ago it was the farm of a Scottish settler, and then of his descendants, and there are still rows of maples planted two hundred years ago. Ghosts live in the barn, but they are old and Scottish and have nothing to do with anyone either Anne or I was ever married to. The land slopes southward,

149

and there's a view of three mountain ranges, a view far better than either of us had at North Farm or South. There are old pastures that had almost gone back to woods, and which together we have reclaimed. That was when Anne learned to drive a tractor. That was the year before Anne got sick.

The one thing we didn't do was build the little house ourselves. Neither of us was a good enough carpenter. We did all the easy parts – say, half the total – and we hired a friend who's a master craftsman to do the hard parts. By the early fall of 1993, we could and did spend nights there.

But this story ends in tragedy. In the spring, Anne had been diagnosed with cancer, the worst kind. Malignant melanoma. What courage and spirit could do, she did. She was still driving nails and finishing a book in September. Between bouts of chemotherapy, that is. But in December she pulled her final disappearing act. Fighting every inch of the way, she died. She was fifty-three. For me all three houses are haunted now. She is a very beautiful ghost.

[1995]

The Guardian Angels of
Tucker Hill Road

❦

Tucker Hill Road in Thetford is one of the Upper Valley's less impressive roads. It's only 2.3 miles long. It doesn't have – or doesn't need – a center line, or any paintwork at all. It averages a little under one streetlight per mile.

And where does this little road go? Well, if you start at the Thetford Center end and drive west to the T-intersection where it meets Vermont 132, you arrive at a place that doesn't even exist. It did. Long ago there was a Mr. Rice, and he had a mill. But Rice's Mills is only a name now.

On the other hand, for about two hundred years this dinky road has offered a series of wonderful views, most of them created by human endeavor, with some assistance from cows and sheep.

Turn around and start back to Thetford Center. On the right-hand side it's all trees for the first mile. Corps of Engineers trees, and not likely to be cut down any time soon. On the left, however, comes one stunning house after another, with wide spaces between them.

First comes one you hardly notice, but you'd love it if you did. It's what passes on Tucker Hill Road as a new house. Rob Hunter, an architect who lives about three miles away, designed it in 1968. Jimmy Banker (born and bred in Thetford) and his crew built in '69. Rob set it at the top end of a little orchard, three hundred feet back from the road, which is why you're more likely to notice apple trees than you are the house. But if you did take a look, I think you'd be enchanted by the way the roofline, with its faintest suggestion of a classic Japanese country house, tucks the house into the landscape.

The next house you can't miss. It's one of the four brick houses that adorn the left-hand side of the road, and it was there well before Mr. Rice. I don't know its exact date, but I'm guessing around 1810. It's in the style called Federal.

The bricks of House 2 are as local as the builder of House 1, though a good deal older. They came from Hezekiah Porter's brick-yard, about two miles away. There's a story, perhaps apocryphal, that Mr. Porter, who also ran a tavern and farmed, made bricks only in his spare time, and so took a couple of years to turn out enough for a new house. What's certain is that all five of the brick houses on Tucker Hill Road (there's one on the right, later) are made of the kind of rosy bricks that modern brickyards seem to have forgotten how to make.

I do know the date of House 3, half a mile down the road. Orange Hubbard built it in 1798. Starting about here, the right-hand side of the road shifts from government trees to privately owned fields.

Another mile and five houses farther on, we come to the third brick house. I don't know its date, either, which is absurd, because I have been living in it since 1963. I'll guess 1805. It's the worst maintained of the five, but still handsome enough so that every year or two someone barges in with a checkbook and tries to buy it. I once considered putting up a sign that would say Not For Sale. But then common sense prevailed. Certain types of buyers would take a sign like that as an interesting challenge.

From here on, there are houses on both sides, though not very many. All but two are old. Keep going. Quite soon, you come to the covered bridge. It's only fairly old and only fairly handsome, though the two great timber arches are worth stopping to look at. But it matters for another reason.

That bridge is one of the four guardian angels that protect

The Guardian Angels of Tucker Hill Road

Tucker Hill Road. The bridge does its share by providing only eleven-foot clearance. Plenty for a loaded hay wagon, but impassable for the larger kinds of trucks. While it exists we are not going to have the kind of clogged-artery Route 4 Disease that poor Woodstock suffers from.

Now you are very near the end of Tucker Hill Road. The fourth brick house, a small, elegant cape, is on your right, and so is a pasture containing three horses and a foal. He's an altogether endearing black foal, who looks like a cross between a seahorse and a chess knight. Not far past this baby and the three grown horses, on the left, is the fifth brick house. It's the biggest and the handsomest of them all. It sits on a little knoll and gazes placidly out on its own hay fields. If I have the facts straight, Hezekiah Porter made a start on this house in 1806, but didn't finish it until 1815. It took a lot of bricks.

There's just one more parcel of land on the road. This parcel happens to have been blessed by the second guardian angel. "Parcel" is a dull word. It's actually an unfenced three-acre field occupying the south corner where Tucker Hill Road runs into the main street of Thetford Center. It's fairly level for a Vermont field, and could easily accommodate a Burger King with attendant parking. Could but won't. It won't because a year before he died in 1997, at age ninety-eight, Charles Hughes of Thetford Center gave that field to the town. He put two conditions on it. One, no buildings, ever. Two, "any hay on the Property shall be cut." Guardian Angel Hughes had in mind a village green, not a thicket.

As I see it, Mr. Hughes has nailed down one end of Tucker Hill Road, and done so with the willing and nonbureaucratic assistance of local government. "The Thetford selectboard is delighted with your generous offer," the board wrote him on November 3, 1997, and it instantly accepted.

So tough luck, Burger King. You'll have to find another spot. You can't build here. And it's no use your zipping down to the other end of Tucker Hill Road, though there's a nice level corner there, too. That end is also nailed down. Barbara Sorenson teaches at Thetford Academy and lives in the house with the pretty roof-line. She owns the corner where Tucker Hill Road connects into Route 132; she owns a few battered foundation stones from Mr. Rice's mill. But she couldn't break up her land and sell a couple of acres to Wendy's even if she wanted to, which she doesn't. The third guardian angel would intervene. This one is the Upper Valley Land Trust, which protects her land and about two hundred other parcels in Vermont and New Hampshire.

Barbara could sell the land; she could even sell it to Wendy's (though our zoning administrator might have something to say about that). It's just that Wendy couldn't put up any structure except with a clear agricultural purpose, make any subdivision, etc., etc. What Wendy's could do would be to graze three or four Angus steers and pick apples.

So both ends of Tucker Hill Road are nailed down. But what about the rest? What protects that? Well, that third angel has just opened her wings over one of the brick houses – the one that Orange Hubbard built in 1798 and that Matt and Martha Wiencke lived in for thirty-seven years, until Matt died last year. Together Matt and Martha planned the protection of their house and their hundred acres that run on both sides of the road. They reserved the right to split off one two-acre building site, in case one of their children should want to build someday.

Don't forget there's a fourth angel. Go on down to the next brick house, mine. Take a good look at Bill Hill, which occupies part of the ninety acres I own on both sides of the road. The Vermont Land Trust protects them all.

In fact, I suggest you go climb Bill Hill. There's a 360-degree view from the top, which you wouldn't expect from so small a hill, plunked down by a river. Like Mr. Hughes, I have a condition. If you do in fact walk across my cow pasture and climb Bill Hill, shut the damn gate behind you. I don't have my little herd of grade Herefords at present, but I do have guest cows each summer and fall. They come from Dick Howard's farm, way up in North Thetford, and they are Holstein heifers. Uncommon pretty against a green background.

The only part I don't like about having guest cows is when they get out. The only time they get out is when someone leaves a gate open. I admit that once, a decade ago, that someone was me.

I can't end without introducing a fifth angel. Ellis Paige, the man who mows most of the fields around Thetford Center and who could probably fix a tractor blindfolded, is one.

I suppose he won't altogether like being called an angel, but it's the truth. Let me show you. You'll recall that Mr. Hughes wanted that three-acre field at the corner to be mowed in perpetuity – or at least while there was any hay to mow. There isn't much. I've mowed that field a few times myself, and except for a strip along the eastern side it's pretty sparse.

Now we come to Ellis's letter. On April 27, 1998, he wrote to the selectmen. Here's what he said: "I'm asking for permission to hay the field that Mr. Hughes has turned over to the town. . . . It will not be for profit but mostly because I would like to see it kept open and cleaned up. I will not charge the Town for the maintenance." Permission was granted that same day. Maybe there are select-angels, too. I think so.

It surely isn't for profit that Ellis mows that little field. You hardly get enough hay to pay for your tractor fuel. It can only be done for love.

Tucker Hill Road, though it still has some vulnerable spots, seems likely to stay a true rural road for some time to come. If we can only find enough angels, every town in the Upper Valley can keep some roads like it.

[2000]

Break & Enter

🍂

I OWN A small camp on a piece of high ground in Caledonia County, Vermont. It's just one room. My wife and I built it together six years ago, with considerable help from a friend who's a professional woodworker. It has a great view – and also a great attraction for people who like to break into remote cabins.

The first break-in occurred before we'd even finished building. Garrett, our woodworker friend, had just made and hung the two doors, front and back. Someone must have taken their installation as a challenge. This someone put his shoulder to the front door with such force that a whole section of the frame snapped off, taking the lock with it. The actual door, I'm glad to say, was unhurt. Garrett had made it out of local butternut and had given it to us as a cabin-warming present. We loved it.

I think the person who broke in, besides enjoying challenges, may have wanted a place to spend the night. He definitely built a fire, helping himself to our firewood, and when he left, he did not bother to shut the now freely swinging door. But he took nothing: not the fine new futon Anne had bought for the sleeping loft, not any of the tools, not even the box of granola bars I kept for midnight hunger pangs.

I was able to put the frame back together well enough so that the lock worked again, and for nearly a year no one broke in. Meanwhile, my wife got sick.

Then the front door got snapped again. This second intruder

did not build a fire or spend the night. But when he left, he did take a souvenir. Even before the doors went on, we had had an eight-panel solar array installed to make electricity, and we had bought a small generator to be the backup if there should be a long stretch of bad weather. This the breaker-in now lugged away.

Fast and angrily I nailed the front door shut, tight into Garrett's butternut boards, using big nails. I wasn't thinking much about neat or careful repairs that fall. My beautiful wife had been diagnosed with the most dreadful of all cancers: malignant melanoma. She fought as hard as a human being can fight, and she lost.

For about a year after her death I couldn't bear to visit the cabin. It had been our private place, where no one went but us. What I did do was write a piece for *Yankee* about her, and about the camp, and about how glorious a late marriage can be ("A House, a Horse, a Hill, and a Husband," July 1995).

When I finally did return, to mow the cabin field, someone had been there before me. Someone had forced the back door open and come in, tracking lots of mud. He hadn't taken anything, though, except a look around. (Mud even in the loft.)

The front door was already nailed shut. I couldn't nail the back one, too, without sealing myself out. The frame was too badly shattered for me to fix. So as a temporary measure I tied the door shut with a piece of baling twine. At least that would keep it from flapping in the wind.

Somehow a month went by without my remembering or doing anything about getting the door fixed. Then, the day before he was to leave for Boston to study music, Anne's nineteen-year-old son, Marek, drove down to the cabin with me. He needed furniture for the tiny apartment he had rented. Since I still didn't (and don't) care to spend nights alone in the cabin, I offered him the futon, and he gladly accepted.

I didn't bother to bring the key. What was there to unlock? One door was tightly nailed shut, the other loosely tied.

We turned off the town road and drove the quarter mile in to the cabin. We walked around back. To my amazement, the back door had healed itself. The baling twine had vanished; the lock was working – and we were locked out. I had never heard of a door healing itself, not even one built by a master woodworker, but this one had. Marek and I sensibly decided to get the futon later, when he would be home for a weekend.

Meanwhile, I had a mystery to solve. The very next day I went back to the cabin, bringing the key. The back door opened easily. And the first thing I saw, as it swung open, was a message. Someone had taken a scrap of clean white-pine board from the scrap pile and had written me a note, using a piece of charcoal from the stove as a pencil.

This is how it read:

SOMEONE BROKE IN YOUR DOOR. I FIXED THE DOOR.

ROB MARCOTTE

And then below his name he had continued:

I LIKE TO COME HERE IN THE WINTER & SIT ON YOUR PORCH. READ THE STORY IN YANKEE & KNEW THIS WAS THE PLACE.

I'm not sure what the opposite of a break-in is. A restoration, maybe? A healing? But I know this. We have them in Vermont.

Postscript: Rob also left his phone number, written in charcoal on the board. Naturally I called that night to thank him. It turned out that he and his girlfriend like to snow-shoe in on late-winter afternoons. Then they sit on the porch and watch the sun set over the three ranges of mountains you can see. I may just give them their own key, in case they should want a fire and some hot boiled cider afterward.

[1999]

Life on Nothing a Week

❦

THETFORD, DECEMBER 5, 1997: Today is the seventh straight day I have spent no money. I haven't used my credit card, either. Only twice all week have I set foot in a store, and neither time did I buy anything. (What did I go in for then? You'll hear.) At no time did anyone else buy stuff for me. It has been an interesting week.

As you might suppose, it started with a dare. I don't often go a week without buying anything – the last time was forty-some years ago, in the trenches, in the Korean war. There was nothing to buy.

I dared myself into this buyless week because Donella Meadows was looking for people who would take a pledge not to buy anything on the day after Thanksgiving. That's the biggest shopping day of the year, the day when we are most fully a consumer society.

"That's not much of a challenge," I said – boastfully, I'm afraid. "I don't buy anything on Thanksgiving Friday anyway. Tell you what. I'll go a week."

Later that day I was telling my friend Terry Osborne about the plan. "What's to stop you from stocking up in advance?" he asked sceptically. I thought fast. "Because I won't know when the week begins until it actually starts. I'll get Donella to pick the starting day, but keep it secret. On the actual day she calls me. I start. If I happen to be out of eggs or beer or gas, tough." Talking with Terry, I also made a policy on invitations. They would be OK provided there were no more than in a normal week.

Donella readily agreed to be my starter. She made the call on

The Day, the Friday after Thanksgiving. She wanted me to begin the same time as everybody else, just go on six days longer.

But I wasn't home to get the message. I was up in Barnet, Vt., where my late wife had a house and where Connie and Marek, my two grown stepchildren, were home for Thanksgiving. They were home for America's favorite shopping day, too. We shopped. Connie and Marek and three of their cousins and I all drove into St. Johnsbury and had brunch at the Northern Lights Cafe – about $40 worth of brunch. Then we shopped like mad. Connie got a fur hat, Suzannah got a new winter jacket, Marek got a gorgeous flannel shirt, I got jeans, and so on. So much for my boast of frugality.

SATURDAY ABOUT 9:30 A.M.: I got home to Thetford. Donella's message was waiting for me. "If you don't get this in time to start on Friday," she said, "then your week begins the instant you do get it."

Ouch. Trouble already. Marek and Connie were passing through Thetford about noon, on their way to Boston. I had said I would take them to lunch at the Thai restaurant in West Lebanon. I like them, and I like to keep my promises. I was horribly tempted to pretend I didn't listen to my messages until after lunch. But no, cheating's no way to start a test. At 9:35 I called them and said we'd have to put off Bangkok Gardens to another time. "Don't worry," said Marek, "we'll take you. We don't do that often enough." Kind of nice to be treated by one's kids.

If this had been an ordinary week, I would probably have gotten gas while I was down there – my farm pickup had just over a quarter of a tank. But, of course one does not get gas without spending money. I promised myself to use my little electric car as exclusively as possible, snow or no snow.

When I got home, I cut wood for a couple of hours. Then at dark I came in, and for the first time in my life did a refrigerator-and-pantry inventory. Depending on how you look at it, I either had appallingly little or astonishingly much.

On the side of the little, I had no meat except an unopened half pound of bacon, and no green vegetables at all. I had maybe a pint of milk, not very fresh, the remnant of an aging half-gallon. No ice cream. Seven eggs. Two bottles of beer.

But on the side of much I had most of a 5-pound bag of sugar, a couple of quarts of my own maple syrup, an unopened 2-pound bag of King Arthur flour and a cup of old flour in the bottom of the flour crock. A 3-pound sack of red potatoes. Also a box of rice, an unopened bag of Goldfish, two big boxes of the Swiss cereal called Familia, plenty of coffee and about six kinds of herbal tea.

Back on the side of little: no bread except two ancient hot dog buns and three almost-as-old hamburger buns in the freezer, along with one very small round loaf of sourdough bread – a 6- or 8-ounce loaf. Also four cherry tomatoes – nice ripe ones, worth two bites apiece. As for nuts, the pantry yielded half of an ancient 12-ounce jar of chunky peanut butter. Of cheese, the tag end of a piece of Cabot cheddar. Finally, my array of canned goods: two cans of Campbell soup, one small can chunk tuna.

And back on the side of much: almost a full pound of butter, eight chocolate chip cookies from Tanyard Farm, three jars of different kinds of honey that people have given me over the last five years, three different kinds of pancake mix (Aunt Jemima buckwheat and two local ones), and countless spices, sauces, pickles and flavorings, mostly very old.

I should pause to explain why I, who like most Americans am truly devoted to ice cream, had none in the freezer, and no frozen veggies or meat, either.

It's because of a bad mistake I made about six years ago. I replaced an aging conventional refrigerator with a much more expensive kind called a Sun Frost. Primary motive: to save on my electric bill. Secondary motive: to cut pollution. The Sun Frost, which is insulated as you have never seen, really does cut electricity usage by 70 percent, as compared to conventional appliances. But the medium-size model that I bought has only one set of controls for both the fridge part and the freezer part.

Predictable result: If the freezer stays cold enough to preserve meat, lettuce begins to freeze in the refrigerator. If the refrigerator stays at proper fridge temperature, ice cream in the freezer gets soupy, and I worry about meat spoiling. So I've wound up using the freezer section – which is pretty small anyway, on account of all the insulation – just for breadstuffs and bags of ground coffee. Someday I will replace the Sun Frost or get it retrofitted.

For dinner last night I had a big bowl of Familia, pouring over it about half my remaining supply of milk. (It's obviously about to turn.) I crunched two of the eight chocolate chip cookies for dessert.

On Sunday morning I always phone my sister in Washington, D.C., at eight A.M. But first I make a cup of coffee using half caf and half decaf. Then I add a good slug of heated milk. This morning the hot milk instantly curdled when I poured it in. I dumped the coffee out, and drank boring black coffee.

My sister and I usually talk for about an hour. So at nine A.M. I began thinking about breakfast. It couldn't be cereal, unless I was prepared to have it with water or with some of the very old powdered milk I found at the back of one cabinet. I was not. I had lots of experience with powdered milk in Korea, and it is truly vile stuff. My seven eggs I wanted to hoard a little.

But there was also an open can of powdered buttermilk that my late wife must have brought down from Barnet around 1992. I would use that, plus flour and baking soda, and I would make biscuits. So I did. Made a dozen biscuits on a teflon cookie sheet.

I have to say they weren't very good. For one thing, most of them stuck, teflon or no teflon. For another, they barely rose. In thickness they resembled cookies, not biscuits. In texture they were a bit leathery. And there was something faintly rancid about the flavor. All the same, I had seven of them for breakfast. Plenty of butter will redeem almost anything. I also employed a little bit of homemade (not by me) strawberry jam left in a jar I bought last June. It nicely wiped out the hint of rancidness.

Naturally I couldn't buy a Sunday paper, and if I'd been going to church I couldn't have put anything in the plate. Couldn't go to the movies or a concert or anything like that. I spent the morning in the woods, and when I came in around 1:30, I had the four cherry tomatoes, the five remaining biscuits, and the last of the strawberry jam. Also one chocolate chip cookie. There are now five left.

Cynthia Taylor, a friend up the road, had heard about my peculiar week, and she called in mid-afternoon to invite me to come have dinner with her and her daughter Zoe, who is my honorary goddaughter. Naturally I went. Not only did Cyn give me a splendid vegetarian stir fry, with salad to follow; we then had chocolate ice cream. Money can't buy love, but it can buy Ben & Jerry's.

Having failed with the biscuits, I decided on Monday morning to try a muffin mix that I picked up three or four years ago on a trip to northern Vermont and never got around to using. And mindful of the faint rancid taste of the biscuits, I decided to skip the powdered buttermilk. I would try the regular powdered milk. And, following instructions on the package, I would put in one of my seven eggs.

I made twelve small muffins. Had three for breakfast, took three more to work in a paper bag to be my lunch, ate four for dinner along with two chocolate chip cookies. Now there are three.

This was a maple-cornmeal mix, and the maple almost wiped out a faint acrid taste that I incline to pin on the long-open powdered milk, though of course it may also have been the long-ground cornmeal. But the maple didn't get every trace. Furthermore, though the mixture was OK, these muffins didn't rise very well, either. Ten of them were enough. I did not save the last two for tomorrow. They went in the compost pail. At bedtime I was still hungry, and I ate both apples.

On Tuesday morning I breakfasted out. My friend Jane Bartlett, after I had taken her to pick up her newly repaired car, took me to breakfast at the new restaurant in the bus station. Eggs! Toast! Bacon! Half-and-half for the coffee! There was a bowl with half a dozen of those little plastic cups of half-and-half on our table, and I thought hard about slipping three or four in my pocket. It wouldn't be buying, just stealing. But no, theft would be against the spirit of the week.

Back in Hanover, I had an errand downtown, and it would have been handy to park and do it. But of course I have no money to put in parking meters – I'm not even carrying a wallet this week, nor is there change in my pockets. I decide the errand can wait.

I go to my office in Steele Hall.

Today I skip lunch, which is a rare thing for me, and I go early and ravenous to Thetford. I eat the entire bag of Goldfish (it's only 6 ounces) plus a piece of bitter chocolate I've been hoarding. What I'd like now is a glass of milk. Milk tastes so good after chocolate. It tastes good after a lot of things. New York City, I hear, is full of people who claim to be lactose-intolerant, but I think the Upper Valley is full of people who really are lactose-dependent. Certainly I am one.

Now I have an idea. I get one of the half-pint cans I will use in maple-sugar season, heat a little syrup on the stove, and can it. Then I hop in the electric car (the truck has about ⅛ of a tank left) and drive the 1½ miles it takes to go the quarter mile into the village, and will continue to until the bridge repairs are finished. I go into the Village Store with the still warm syrup, and explain that I am in the middle of a week where I can't spend money. "Could I possibly swap syrup for milk?" I ask. Bev says yes. I dance off to the electric car with a half gallon of good honest whole milk, and as soon as I get home I sit down to a big bowl of Familia, drenched in milk.

I suppose you think that means I skipped supper? Not likely. Having this nice fresh milk, and having decided to commit half of the little dribble of canola oil at the bottom of an almost empty bottle, I make another batch of biscuits. I double the amount of baking soda Ms. Rombauer calls for (God knows how long the baking soda has been open), and I use a muffin tin, not that miserable teflon-coated baking sheet. Using my right forefinger, I quickly rub a little butter in each muffin hollow. Then I drop the batter in carefully, filling each hollow just so.

I turn out to have enough batter for nearly ten muffo-bisks. They are excellent. I eat six of them for a late supper, the last one swimming in maple syrup. I wash it down with a glass of milk. God bless the Village Store! I wonder how my proposal would have fared at Grand Union, or Wal-Mart.

The next day is Wednesday, and I don't have to be in Hanover until 4:30 P.M., when I am one of several people doing a signing at the Dartmouth Bookstore. (That will be my second and last time in a store this week.)

I eat the other four of my high-quality biscuits for breakfast, and I feel so confident I even soft-boil an egg. There is hot milk for coffee.

You may be getting bored with reading about biscuits, but I was not yet bored with making them. Around noon on Wednesday I used the last of my flour to make twelve large muffo-bisks. Delicious, my best yet. At lunch I employed the three biggest as vehicles for a taste test: I wanted to compare the three jars of honey that have been kicking around my pantry for so long. One is a jar of Langnese honey from Europe, so old it's totally crystallized, even granulated. One is raspberry honey from Oregon, and one is fireweed honey made by my brother-in-law in Washington state. All are so good that I can't possibly rank them. The granulated Langnese can be eaten with a spoon – it has a lovely texture. The other two, the liquid honeys, make a good biscuit great.

My one conclusion: standard clover honey is boring compared to any of the three.

That afternoon I parked at a meter while I was signing at the bookstore, and didn't get caught. Dinner I had at Jane's, and tasted meat. But Thursday I was back to muffo-bisks. I work in the environmental studies program at the college, and we have a brown-bag lunch every Thursday during terms. My brown bag contained four muffo-bisks, the jar of Langnese (much tidier than the two liquid honeys) and a spoon.

I will admit that I was now ready for a change. That evening I dined on rice, plentifully seasoned with teriyaki sauce from a bottle that probably dates from the late 1980s.

And so it went until the week ended at 9:30 on Saturday morning, December sixth. No serious problems. Oh, there were a couple of inconveniences. For example, I have a recurrent fungus-on-the-feet infection that I picked up in Korea. It chose this week to recur. It would have been handy to swing by the Thetford Pharmacy and get some ointment. But a few days really doesn't matter. No toes will drop off.

Not only were there no serious problems, I ended the week with surpluses. The electric car did so well on snow and ice that my four-wheel-drive pickup still had a sixteenth of a tank left, maybe more. I still had one chocolate chip left, and three eggs. But that's nothing. I also had the untouched bag of potatoes and more rice. If need be, I could probably barter syrup for more flour. I could easily do another week, maybe even a month. Sometime in the summer, when green vegetables come from gardens, I may just try it.

What is the point of this week? Well, there are two points. One is the pleasurable challenge of being thrown on one's own resources. Of being independent, self-sustaining. It is extremely handy going to Dan & Whit's and Wing's and the Co-op and the Village Store, and I would not wish to give any of them up. But occasionally it seems too easy: just picking up frozen everything or a lot of prepared foods. And then it's time to make a few muffo-bisks.

The other reason, of course, is to practice consuming less. Not necessarily less food (I've probably gained a pound or two this almost saladless week). Less energy, fewer jackets, fur hats, extra jeans, exercise machines, kitchen gadgets, disposable cameras, plastic spoons. Of most kinds of possessions we Americans have more than we know how to use. Which is one reason we consume a quarter of the energy and produce a quarter of the pollution in the world.

It won't be easy to change our ways. High consumption is one of our oldest and most powerful traditions. It is by no means an entirely bad tradition, either, being closely entwined with the whole concept of generosity, of knowing that it is blessed to give, and if not damned at least very unattractive to be a tight-wad and a scrimper.

But clearly as a society we have carried consumption too far. Long ago we turned shopping into the most expensive – and, yes, deadly – of all sports. Hunting in the woods got replaced by hunting in the store. One of the funniest and most historic examples occurs in a novella Henry James wrote in the 1870s. It's called *An International Episode*, and it takes place in Newport, Rhode Island.

Two visiting Englishmen, Mr. Percy Beaumont and young Lord Lambeth, are staying with the wealthy Westgate family. Its principal members present in Newport are the beautiful Mrs. Westgate, age thirty, and equally attractive younger sister Bessie Alden, age twenty.

Both Englishmen are soon put to work assisting these two in their shopping expeditions. When Percy Beaumont mildly protests, Mrs. Westgate explains that such expeditions are not optional.

"An American woman who respects herself," says Mrs. Westgate, fixing her beautiful eyes on him, "must buy something every day of her life." Pretty Bessie then demonstrates a remarkably high level of self-respect. Taking Lord Lambeth along to mind the parcels, Bessie cruises seventeen shops in one morning. They weren't using the term "mall" yet (not in its present sense, anyway), but clearly Bessie was doing a mall crawl.

That was in 1874. Now, a century and a quarter later, Bessie's descendants, male and female, need a new way to achieve self-respect. For starters, they might try a low-consumption, do-it-yourself week now and then.

[1997]

167

Farewell to a Thetford Farm

℟·℣

ABOUT THREE MONTHS AGO my wife and I made a major decision. We decided to sell our farm. We did this knowing it would be
painful for both of us. I have sometimes thought it was a little
extra painful for me, just because I have known the eighty-five
acres of our farm so long and so well.

I've been living on the farm for the last forty-one years. I love
every acre. I also love the brick farmhouse, which I believe to be
about 190 years old, and which I know for sure was built of local
Thetford bricks. I have plenty of love left for the two barns, one
early nineteenth century and one late nineteenth century.

The handsome old brick farmhouse is where I have lived almost
my entire adult life, and I have no wish to move. It's also where my
younger daughter, Amy, came into the world thirty-some years ago;
she has no wish for me to get rid of the house where she was born.

What about Sara, my tall, beautiful, book-addicted wife? She has
lived at the farm only the three years since we got married, which
is about one-fourteenth as long as I have. But she has crowded an
amazing amount of rural know-how into those three years. For
now, I'll just mention that she not only runs a chain saw, she is so
good at starting a reluctant Jonsered or Stihl that I don't even try
anymore. I pass my saw over to Sara. She starts it, lets it run for half
a minute, cuts the engine, and hands it back. Those big steel teeth
are perfectly motionless.

Or consider tractors. We have a medium-sized one, a John

Deere 1050. It took Sara about twenty minutes to get the hang of
mowing our field with it. We are both so fond of mowing that we
have had to make almost a formal rule: mowing is to be shared
50-50. And, oh yes, the quantity of land devoted to perennials has
shot up in the past three years.

But I'm getting ahead of my story. I haven't introduced the
farm yet. I will do so now. And I'll start with the location.

There are two covered bridges in the town of Thetford. There
used to be five, but that was before Union Village Dam got built.
One of the survivors is on Tucker Hill Road, just outside the village
of Thetford Center. Coming almost up to the bridge is a fenced
pasture for cows, though there are no cows in it now. Just outside
the pasture and even closer to the bridge there's a small sugarhouse with a freshly painted red door and two cracked windows.
It was last used to make syrup in April of 2004.

With a great deal of help, I built that little sugarhouse way back
in 1969. It pleased me greatly that I was able to put recycled siding
on, using hundred-year-old boards that I had bought – at a very
modest price – from a man who was taking down an old barn,
I think, in Sharon. Don't worry about my putting clear pine board
to so humble a use. They were – still are – humble boards. Lots of
big nail holes. And the man in Sharon must have worked fast.
Almost every board had scars. I think they were being given a second chance when he sold them to me.

Wooden bridge with great curving timbers, old-fashioned sap
buckets on the nearby maples (March and April only), well-
fenced pasture – this part of the farm has an enormous quaintness
quotient. And people often do stop and take pictures. I have nothing against quaintness. In fact I rather like it, as long as it's unself-
conscious. But it's not what I love the place for. I love the place we
sometimes call Two of Everything farm for about twenty reasons,
maybe twenty-five. For example, I dote on the old brick farmhouse
– and it's a three-way dote. First, I love the look and the feel of the
old bricks. They're softer than new bricks, and have a better color.
Second, I like it that they are a local product, made in Thetford

some time around 1810. Most of all I like it that the house has style. It's a rural adaptation of the urban architecture known as Federal, and it's a delight. My heart does a little skip every time I come in.

Two more reasons. I love the view from the kitchen window. It's a rolling pasture with trees behind. Other than the barbwire fence, which is inconspicuous, there is no trace of the works of man. And of course I delight in the grander view (360 degrees) to be had if one climbs Bill Hill, the Farm's in-house mini-mountain. From the top you can see about five miles.

So. Now we have identified five of the reasons why the little farm is so appealing. They have either been small ones, or at most medium-sized. Are you ready for a biggie? Good. The thing that delights me most is that the farm really is a farm . . . about 10 percent of the time. It does produce a little food every year, and most years a little fuel as well.

I'll skip over some early attempts to market produce, and go straight down to that little sugarhouse near the covered bridge. I first made maple syrup there in 1970, and I made a paltry four gallons. A few years later, there came a March and April of perfect sugaring weather, and I happened to be sugaring with an energetic partner that year. We made fifty-seven gallons. That's a quarter ton of syrup.

At one time I thought I might turn most of my share into maple sugar, which could then be used as a general sweetening agent. The point would have been to supply a local organic product to serve as an alternative to cane sugar. But it would never have worked. Apart from sugar made from fancy grade syrup, there is just too much flavor.

As to fuel, in our best year we cut and split a little over twenty cords of firewood, double what we did any other year. That meant five cords for the wood stoves in the old brick farmhouse, and fifteen cords to sell. Our best customers were nearly all Dartmouth professors.

Another thing I love about the farm is that there's always something to do. I'll even say something that needs doing. For about thirty years my favorite something has been the care and repair of stone walls, plus the very occasional construction of a new piece of wall.

Repair is something you can do alone – though it's less fun that way – but a from-scratch piece of new wall really calls for a two-person crew. In my case, the other person was nearly always a local physician who practiced medicine part-time, precisely so that he would have time for things like wall work. His name is Andy.

In those early years I believed that to make a decent wall from scratch you needed to excavate a trench where you intended the wall to go. Then (so I thought) you had to put a lot of your good rocks in the trench, as a foundation for the wall-to-be.

The good news is that you don't have to do any such thing. For a mortared wall, yes. But for a regular stone wall, optional. You can lay the first course right on the ground if you like. Though I do recommend taking a few minutes to make sure that the first course is level. It just isn't hiding in a trench, where no one will ever see it.

For about ten years, mostly in the '90s, Andy and I traded stone work on Wednesdays. Mornings only, and just every other week. We each had a wall maybe one hundred yards long and badly in need of repair. Besides the replacement of whatever big stuff had fallen off the wall during the last hundred years, we inserted many chinkers. They are small flat rocks, seldom weighing as much as a pound, and they are wonderful stabilizers.

We finished Andy's wall two years ago. There's still plenty left to do on mine – it's a big as well as a long wall – but it's doubtful that we'll get to much of it.

I have developed a remarkably unpleasant version of Parkinson's disease. One of the unpleasant things it does – one of the minor ones – is make it impossible for me to lift heavy rocks. Heavy anythings, actually. Thank God for chinkers.

There are many other things to do at Two of Everything Farm. Cider-making, for example. For about a century it has been possible to buy a one-bushel cider press. Then, when you climb up Bill

Hill with a picnic lunch and maybe a pair of binoculars, you can also bring a jug of fresh-pressed cider. If any children are present, you can be pretty sure they will have ground the apples, done the pressing, and now are glad to carry "their" cider up the steep side of Bill Hill.

Then there are cows. There have been cows at Two of Everything Farm for about thirty of the forty-one years that I was privileged to live there myself. Mostly they were guest cows; they came in June and stayed until October or maybe early November and day by day trimmed up whichever pasture they were in. For a few years I had a tiny herd of beef cattle, and that was the period when I felt most at home on the farm. I love cows – for their warm sweet breath, their sweet dispositions (I could write a whole piece about the week when I had to give two shots a day to a sick heifer) and their insatiable curiosity. Once when my sister Bee was visiting, and we were making cider (kids at school), the six guest heifers we had that year were lined up like the audience at a play. They stayed neatly in a row along the wire fence, and did all but clap. I rewarded them with all the pomace from three one-bushel pressings.

As I move into exile – and that is how I see leaving the farm, the maple trees, the cattle, the wild turkeys – I am very clear that assisted living comes at a price.

[2004]

BIBLIOGRAPHY

A Passport Secretly Green. New York: St. Martin's, 1961.

Dr. Bowdler's Legacy: A History of Expurgated Books. Hanover, NH: University Press of New England, 1969.

Amateur Sugar Maker. Hanover, NH: University Press of New England, 1972.

Vermont in All Weathers. New York: Penguin Group (USA), Ltd, 1973. (or Viking?)

The Adventures of Jonathan Corncob: Loyal American Refugee, Written by Himself. (Editor.) Boston: David R. Godine, 1976.

Giving Up the Gun: Japan's Reversion to the Sword, 1543–1879. Boston: David R. Godine, 1978.

First Person Rural: Essays of a Sometime Farmer. (Illustrated by Stephen Harvard.) Boston: David R. Godine, 1978.

Second Person Rural: More Essays of a Sometime Farmer. (Illustrated by F. Allyn Massey.) Boston: David R. Godine, 1980.

Third Person Rural: Further Essays of a Sometime Farmer. (Woodcuts by Robin Brickman.) Boston: David R. Godine, 1983.

Mills and Factories of New England. (Co-author with Kenneth Breisch. Photographs by Serge Hambourg. Captions by Kenneth Breisch.) New York: Harry N. Abrams, 1988.

A Reader's Delight. Hanover, NH: University Press of New England, 1988.

Last Person Rural: Essays by Noel Perrin. (Illustrated by Michael McCurdy.) Boston: David R. Godine, 1991.

A Noel Perrin Sampler. Hanover, NH: University Press of New England, 1991.

Solo: Life with an Electric Car. New York: W.W. Norton and Co., Inc., 1992.

A Child's Delight. Hanover, NH: University Press of New England, 1997.

ACKNOWLEDGMENTS

No book – especially a collection like this, which has pulled material together from a number of different sources – can exist without the help of many people. I want to acknowledge those people here and express my deep gratitude.

The book would never have happened without David Godine. His idea inspired the project. He proposed the idea to Nardi Campion, and Nardi graciously passed the proposal on to me. Since then, David's suggestions have helped to shape the collection, and his sharp editorial eye has smoothed out the rough spots in my prose.

Thanks also to everyone at David R. Godine, Publisher, who helped with the preparation and publication of the book.

Sara Coburn assisted me throughout the project, answering my questions as they arose, providing materials when I needed them. I've been honored by her patient support.

The same is true for Elisabeth Perrin and Amy Perrin Haque Joy. I hope this collection honors their dad the way he deserves to be honored.

Nat Tripp provided manuscripts and photographs. Jon Gilbert Fox, Medora Hebert, Annemarie Hoffmeister, Nancy Hunnicutt, Reeve Lindbergh, and Laurel Stavis all contributed or located or tried to locate photographs.

Don Mahler tracked down some of Ned's uncollected essays and patiently tracked them down again when I lost them. I'm grateful for his friendship.

Finally, the following people all contributed in some way to the Foreword: MK Beach, Bill Cook, Nancy Crumbine, Jayne Demakos, John Elder, Andy Friedland, Cynthia Huntington, Barbara Krieger, Pat Parenteau, Richard Brooks, Jim Schley, Scott Stokoe, and Tom Slayton. The piece is immeasurably better because of their attention and help.

Thank you all.

PERMISSIONS

Great care has been taken to trace all owners of copyright material included in this book If any have been omitted or overlooked, acknowledgment will gladly be made in future printings.

The essays in *Best Person Rural* originally appeared in the following publications:

Boston Magazine: "Vermont Silences"; *Country Journal*: "Sugaring on $15 a Year," "In Search of the Perfect Fence Post," "Best Little Woods Tool Going," "Maple Recipes for Simpletons," "The Rural Immigration Law," "Birth in the Pasture," "How to Farm Badly (And Why You Should)" "A Truck With Pull," and "My Farm Is Safe Forever"; *Harrowsmith*: "The Lesson of the Bolt Weevils"; *Inquiry*: "Nuclear Disobedience"; *The Los Angeles Times*: "Two Letters to Los Angeles"; *Upcountry*: "The Year We Really Heated with Wood"; *Valley News*: "The Guardian Angels of Tucker Hill Road," "Life on Nothing a Week," and "Farewell to a Thetford Farm"; *Vermont Life*: "Grooming Bill Hill" and "The Two Faces of Vermont"; *Yankee*: "Break & Enter." "Jan Lincklaen's Vermont" first appeared in *First Person Rural*. "The Soul of New England" first appeared in *Last Person Rural*; "A Vermont Christmas" is reprinted by permission of Land's End, Inc.

Photograph on page ii reprinted courtesy of Medora Hebert; photographs on pages 33 and 67 by Richard W. Brown; photograph on page 103 by Robert Pope, reprinted by permission of *Valley News*; photograph on page 141 by Jennifer Hauck, preprinted by permission of *Valley News*. Our thanks to the staff of *Vermont Life* for their assistance in locating some of the photographs used in this volume.